VOLUME SEVENTY NINE

ADVANCES IN GENETICS

ADVANCES IN GENETICS, VOLUME 79

VOLUME SEVENTY NINE

ADVANCES IN GENETICS

Edited by

Theodore Friedmann
Department of Pediatrics,
University of California at San Diego,
School of Medicine, CA, USA

Jay C. Dunlap
The Geisel School of Medicine at Dartmouth,
Hanover, NH, USA

Stephen F. Goodwin
Department of Physiology, Anatomy and Genetics,
University of Oxford,
Oxford, United Kingdom

Amsterdam • Boston • Heidelberg • London
New York • Oxford Paris • San Diego
San Francisco • Singapore • Sydney • Tokyo
Academic Press is an imprint of Elsevier

Academic Press is an imprint of Elsevier
225 Wyman Street, Waltham, MA 02451, USA
525 B Street, Suite 1900, San Diego, CA 92101-4495, USA
Radarweg 29, PO Box 211, 1000 AE Amsterdam, The Netherlands
The Boulevard, Langford Lane, Kidlington, Oxford, OX51GB, UK
32, Jamestown Road, London NW1 7BY, UK

First edition 2012

ISBN: 978-0-12-394395-8
ISSN: 0065-2660

For information on all Academic Press publications
visit our website at store.elsevier.com

Printed and bound in USA
12 13 10 9 8 7 6 5 4 3 2 1

CONTENTS

CONTRIBUTORS

Numbers in parentheses indicate the pages on which the authors' contributions begin.

Sarika Chaudhary (87)
UCSF Helen Diller Family Comprehensive Cancer Center, San Francisco, CA, USA.

Gary B. Fogel (1)
Natural Selection, Inc., San Diego, CA, USA.

Andrew Hinton (1)
Pediatric Diabetes Research Center, University of California, San Diego, La Jolla, CA, USA.

Shaun Hunter (1)
Pediatric Diabetes Research Center, University of California, San Diego, La Jolla, CA, USA.

Charles C. King (1)
Pediatric Diabetes Research Center, University of California, San Diego, La Jolla, CA, USA.

Shashi Kumar (87)
International Center for Genetic Engineering and Biotechnology, Aruna Asaf Ali Marg, New Delhi, India 110067.

Mark S. LeDoux (35)
Department of Neurology, University of Tennessee Health Science Center, Memphis, TN, USA; and Department of Anatomy and Neurobiology, University of Tennessee Health Science Center, Memphis, TN, USA.

Kamiya Mehla (87)
Eppley Institute for Research in Cancer and Allied Diseases, University of Nebraska Medical Center, Omaha, NE, USA.

Jiang Qian (123)
The Oncology Center, Johns Hopkins University School of Medicine, Baltimore, MD, USA; and Department of Ophthalmology, Johns Hopkins University School of Medicine, Baltimore, MD, USA.

Gloria Reyes (1)
Department of Pharmacology, University of California, San Diego, La Jolla, CA, USA.

Devi Singh (87)
Molecular Biology Laboratory, Department of Genetics and Plant Breeding, Sardar Vallabhbhai Patel University of Agriculture and Technology, Meerut, UP, India.

Pankaj K. Singh (87)
Eppley Institute for Research in Cancer and Allied Diseases, University of Nebraska Medical Center, Omaha, NE, USA.

Heng Zhu (123)
Department of Pharmacology and Molecular Sciences, Johns Hopkins University School of Medicine, Baltimore, MD, USA; The High-Throughput Biology Center, Johns Hopkins University School of Medicine, Baltimore, MD, USA; and The Oncology Center, Johns Hopkins University School of Medicine, Baltimore, MD, USA.

CHAPTER ONE

From Pluripotency to Islets: miRNAs as Critical Regulators of Human Cellular Differentiation

Andrew Hinton[*,1] **Shaun Hunter**[*,1] **Gloria Reyes**[†,1] **Gary B. Fogel**[‡] **and Charles C. King**[*,a]

[*]Pediatric Diabetes Research Center, University of California, San Diego, La Jolla, CA, USA
[†]Department of Pharmacology, University of California, San Diego, La Jolla, CA, USA
[‡]Natural Selection, Inc., San Diego, CA, USA
[a]Correspondence should be addressed to: http://chking@ucsd.edu/

Contents

---

[1] These authors contributed equally to this work.

Advances in Genetics, Volume 79
ISSN 0065-2660,
http://dx.doi.org/10.1016/B978-0-12-394395-8.00001-3

Abstract

MicroRNAs (miRNAs) actively regulate differentiation as pluripotent cells become cells of pancreatic endocrine lineage, including insulin-producing β cells. The process is dynamic; some miRNAs help maintain pluripotency, while others drive cell fate decisions. Here, we survey the current literature and describe the biological role of selected miRNAs in maintenance of both mouse and human embryonic stem cell (ESC) pluripotency. Subsequently, we review the increasing evidence that miRNAs act at selected points in differentiation to regulate decisions about early cell fate (definitive endoderm and mesoderm), formation of pancreatic precursor cells, endocrine cell function, as well as epithelial to mesenchymal transition.

I. INTRODUCTION

Gene expression is regulated by transcription factors, epigenetic factors, and small RNA molecules, including microRNAs (miRNAs). The mechanism of miRNA action on protein expression has provided insight on how gene regulatory networks control cell physiology, development, and disease. miRNAs are small noncoding RNAs that regulate post-transcriptional gene networks and function in a manner analogous to transcription factors (Bartel, 2009). Mature miRNAs are partially complementary to one or more messenger RNA (mRNA) molecules and function primarily to downregulate gene expression (Bagga and Pasquinelli, 2006; Massirer and Pasquinelli, 2006; Suh *et al.*, 2004). Biogenesis of a canonical miRNA begins with synthesis of a primary transcript (pri-miRNA) in the nucleus from which one or more stem-loops are released by Drosha/DGCR8 cleavage (Bartel, 2004). The resulting hairpin RNA is transported to the cytoplasm by Exportin-5 and processed into a miRNA-5p:miRNA-3p duplex by Dicer. One of the mature ~22 nucleotide (nt) miRNA strands of the duplex is then preferentially incorporated into the RNA-induced silencing complex and serves as a sequence-specific guide for the negative regulation of target mRNAs (Fig. 1.1). There are currently over a thousand experimentally validated miRNA genes in the human genome, predicted to target many thousands of mRNAs. Although the number of defined miRNA targets is relatively small, the number is certain to grow in the future. This is because a single miRNA family can target multiple mRNAs within a given cell, distributing their effect to a posttranscriptional network

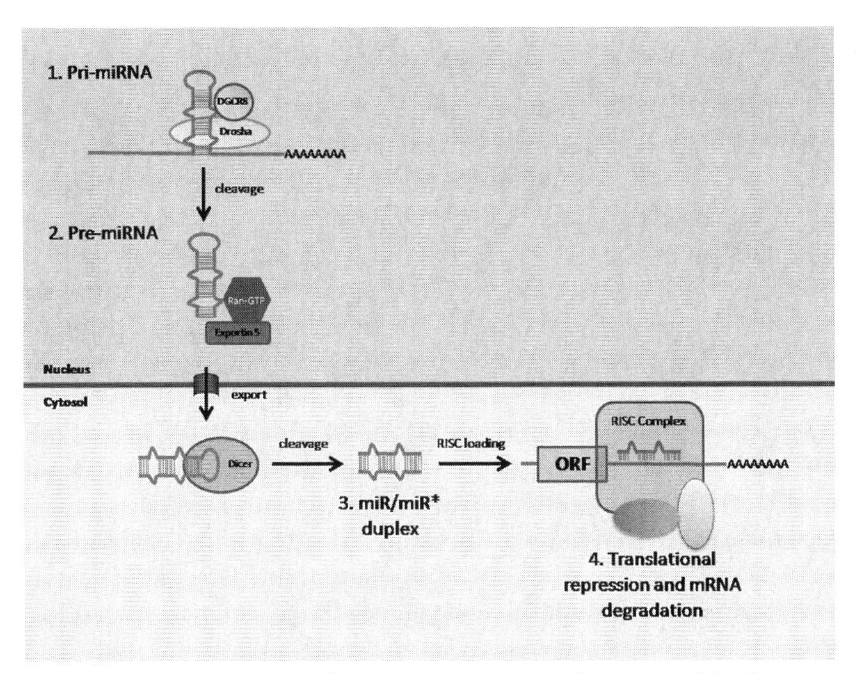

Figure 1.1 Schematic diagram of miRNA processing. For color version of this figure, the reader is referred to the online version of this book.

or pathway. The importance of miRNA activity during mammalian development has been established by deleting genes necessary for global miRNA biogenesis. For example, in mouse embryonic stem cells (ESC)s, deletion of DGCR8 resulted in impaired ability of the mouse ESCs to differentiate (Suh *et al.*, 2010). Similarly, targeted deletion of Dicer in mouse pancreatic precursors resulted in severe deficiency of endocrine tissue formation (Lynn *et al.*, 2007).

ESCs are defined by their self-renewal and ability to differentiate into any cell type and are therefore a potential source for therapeutically useful cells, such as the generation of pancreatic β-cells from hESCs for the treatment for type 1 diabetes. However, a clearer understanding of the molecular mechanisms and pathways, which regulate cell differentiation and lineage specification, is required for directed differentiation into cells of pancreatic lineage. In ESCs, this includes miRNAs which play a mayor role in the maintenance of pluripotency, proliferation (Mallanna and Rizzino, 2010; Melton and Blelloch, 2010; Wang *et al.*, 2008) and lineage specification, including endocrine pancreas (Hinton *et al.*, 2010; Ivey *et al.*, 2008; Poy *et al.*, 2009; Rosa *et al.*, 2009).

II. miRNAs AND EMBRYONIC STEM CELLS

In ESCs, miRNAs play a role in the maintenance of pluripotency and proliferation (Mallanna and Rizzino, 2010; Melton and Blelloch, 2010; Wang *et al.*, 2008), as well as differentiation (Hohjoh and Fukushima, 2007; Lakshmipathy *et al.*, 2007; Suh *et al.*, 2004). Results from our laboratory and others confirm differential and temporal expression of several miRNAs in differentiating human ESCs (Bar *et al.*, 2008; Hinton *et al.*, 2010; Kim *et al.*, 2011a). miRNAs regulate lineage specification for all three primary lineages (ectoderm, mesoderm, and endoderm), and specific miRNAs have also been shown to influence specific cell fates in later stages of development (Hinton *et al.*, 2010; Ivey *et al.*, 2008; Poy *et al.*, 2009; Rosa *et al.*, 2009). For example, misexpression of miR-1 and miR-133 in ESCs is known to cause enhancement of mesoderm at the expense of endoderm and ectoderm formation (Ivey *et al.*, 2008; Takaya *et al.*, 2009). Analogous studies have shown that miR-124 and miR-9 are involved in ectoderm formation during ESC differentiation (Krichevsky *et al.*, 2006). From these studies, it is clear that establishing molecular profiles of mRNA and miRNA expression at each step in developmental pathways will reveal the function of additional miRNAs during embryonic development.

miRNAs are likely to be key components of cellular differentiation. A central focus of laboratories working for cell replacement therapies in type 1 diabetes is to develop glucose-sensitive insulin-secreting cells from pluripotent ESC by mimicking stages of *in vivo* cell differentiation. Therefore, this review will focus on the role of miRNAs in (1) the maintenance of human ESC pluripotency, (2) the formation of pancreatic precursors and endocrine cells, and (3) the function of mature human islets. Increased knowledge about the role of miRNAs in all three of these processes will allow us to transform our ability to mimic cell differentiation *in vitro* and *in vivo*.

III. miRNAs AND EMBRYONIC STEM CELL PLURIPOTENCY

Pluripotency of stem cells is characterized by the ability to self-renew and to differentiate into all three of the primary germ layers. The factors that lead to self-renewal or differentiation depend on autocrine and paracrine signals from the external environment, cell density, and biophysical parameters of the internal environment. Three transcription factors, OCT4, SOX2, and Nanog, are considered master regulators that operate in conjunction with miRNAs and epigenetic mechanisms to maintain an undifferentiated state by silencing developmental genes and promoting pluripotency factors.

To maintain pluripotency, developmental genes must be actively silenced in mouse and human ESCs. Additionally, genes involved in early differentiation must also be silenced but poised to be rapidly activated in response to appropriate cues. Once the differentiation cues are received by the cell, pluripotency genes need to be rapidly shut down and early differentiation genes must be turned on to begin the transition into a differentiated cell. miRNAs can provide support for the dynamics needed to facilitate this change by acting rapidly to suppress the expression of genes that regulate differentiation. Three independent studies in mouse ESCs disrupted global miRNA biogenesis by knocking out either DCGR8 (Wang *et al.*, 2007) or Dicer (Kanellopoulou *et al.*, 2005; Murchison *et al.*, 2005). Results from these studies demonstrated that without mature miRNAs, mouse ESCs displayed decreased proliferation rates primarily as a result of an accumulation of cells in the G1 phase of the cell cycle. Furthermore, DGCR8-null mouse ESCs and Dicer-null mouse ESCs exhibited an inability to differentiate completely, exhibiting a lack of differentiation markers, and the failure to silence OCT4. Moreover, a deficiency in miRNA biogenesis was lethal during early embryogenesis. Clearly, miRNAs are essential for proper early embryonic development. Investigating the role of miRNAs in ESC proliferation, Wang *et al.* (2008) transfected DGCR8-null mouse ESCs with miRNA mimics and discovered a class of miRNAs that could rescue the defect in proliferation, which were then described as ES cell specific cell cycle (ESCC)–regulating miRNAs.

Highlighting the key role of OCT4, SOX2, and Nanog in maintaining pluripotency, ectopic expression of these transcription factors in varied combinations with other genes, including c-Myc, KLF4, and Lin28, has driven reprogramming of somatic cells into induced pluripotent stem cells (iPSCs) that resemble ESCs (Takahashi *et al.*, 2007). Ectopic expression of specific miRNAs has been shown to facilitate reprogramming by transcription factors and more recently to reprogram somatic cells independently of ectopic transcription factors. Several of these ESCC miRNAs share a common seed sequence, and the roles of selected miRNAs in pluripotency and reprogramming will be discussed below.

A. miR-302/367 Cluster (miR-302a, miR-302b, miR-302c, miR-302d, miR-367)

The miR-302 cluster expresses a primary transcript that contains five stem-loops, including miR-302a–d and miR-367. The mature miRNA sequences are highly conserved in mammals, and miR-302a–d share a common seed sequence, while miR-367 does not. Early miRNA expression profiling studies reported that miRNAs from the miR-302/367 cluster are highly expressed in both mouse and human ESCs and that these miRNAs were downregulated during undirected differentiation (Bar *et al.*, 2008; Houbaviy *et al.*, 2003; Morin *et al.*, 2008; Suh *et al.*, 2004). Moreover, expression of the miR-302/367 cluster is dependent on the activity of transcription factors OCT4, SOX2, and Nanog. All three transcription factors bind to the predicted promoter region of the miR-302/367 cluster, suggesting that these factors are transcriptionally, and possibly synergistically, activating the miR-302/367 primary transcript in pluripotent cells (Card *et al.*, 2008). Additional DNA binding studies have shown that the promoter is co-occupied by Nanog and TCF3 as well as SOX2 and OCT4 (Barroso-del Jesus *et al.*, 2009; Marson *et al.*, 2008).

A subset of the mature miRNAs from the miR-302/367 cluster are part of the ESCC miRNAs described by Wang *et al.* (2008), and the abundance of miRNAs in this group with a common seed sequence suggests that they regulate similar targets that help maintain pluripotency. Ectopic expression of miR-302b, miR-302c, and miR-302d rescues proliferation defects of DGCR8-null mouse ESCs. The miR-302 cluster was shown also to promote proliferation in human ESCs. miR-302a, miR-302b, miR-302c, and miR-302d all repressed, but did not completely abolish, the posttranscriptional expression of cyclin D1. Inhibition of miRNA expression resulted in upregulation of cyclin D1 protein in human ESCs (Card *et al.*, 2008).

Maintenance of pluripotency also depends on signal transduction through Activin/transforming growth factor-beta (TGF-β) family members (Beattie *et al.*, 2005; James *et al.*, 2005), and miR-302 participates in promoting this pathway by inhibiting Lefty, a TGF-β/Nodal/Activin inhibitor (Barroso-del Jesus *et al.*, 2009; Rosa and Brivanlou, 2009). Differentiation into a neurectodermal lineage is considered by most to be the default pathway if ESC differentiation occurs in the absence of cues that direct cell fate, so it is not surprising that maintenance of pluripotency requires tight regulation of genes involved in early ectoderm specification, including NR2F2. The observation that OCT4 and miR-302 cooperate to directly repress NR2F2 clearly indicates the multilevel integration of miRNAs with proteins in maintenance of pluripotency. As differentiation proceeds, miR-302 levels go down in ectodermal lineages, *NR2F2* expression increases and acts to directly repress *OCT4* expression (Rosa and Brivanlou, 2011). The miR-302/367 cluster was also shown in a separate study to enhance suppression of neural differentiation by targeting inhibitors of the BMP pathway, including TOB2, DAZAP2, and SLAIN1 (Lipchina *et al.*, 2011). Ectopic expression of the entire miR-302/367 cluster was used to reprogram both mouse and human somatic cells to iPSCs independently of exogenous transcription factors (Anokye-Danso *et al.*, 2011). Mouse cell reprogramming required a histone deacetylase inhibitor additionally, while human cell reprogramming was accomplished solely through modulated miRNA expression. Although all miRNAs in the cluster appear to be essential for complete reprogramming, the roles of individual miRNAs have not been fully elucidated. A different recent study that used miR-302b and miR-372 to enhance transcription factor–mediated reprogramming in human fibroblasts identified several targets (Subramanyam *et al.*, 2011). Among the identified targets were proteins involved in cell cycle regulation (CDKN1A, RBL2, CDC2L6), epigenetic regulation (MECP2, MBD2, SMARCC2), and epithelial-to-mesenchymal transition (EMT) (RHOC, TGFBR2).

The mesenchymal-to-epithelial transition (MET) has been shown to be necessary for reprogramming from fibroblasts to pluripotent cells (Li *et al.*, 2010; Samavarchi-Tehrani *et al.*, 2010). However it is yet not clear whether miRNAs regulate these processes. Consistent with their role in reprogramming, miR-302b and miR-372 expression downregulates several genes involved in EMT in addition to RHOC and TGFBR2, which were validated as targets by luciferase reporter methods. These expression changes were independent of transcription factors (ZEB1 and FN1) or within the context of reprogramming (ZEB2). miR-302b and miR-372 also inhibited TGF-β–induced EMT in keratinocytes (Subramanyam *et al.*, 2011). Gastrulation in embryos involves an EMT; therefore, it seems likely that repression of this process is necessary for maintenance of the undifferentiated state. However, activin signaling at low levels is necessary to maintain ESCs in culture (Beattie *et al.*, 2005), and expression of miR-302d and miR-367 is induced by Activin A (Tsai *et al.*, 2010). However, activin/nodal is also necessary at higher levels to induce an EMT during mesendoderm formation (D'Amour *et al.*, 2005; Kubo *et al.*, 2004). Furthermore, the miR-302/367 and miR-371-373 clusters either persist or are elevated in definitive endoderm during hESC differentiation (Hinton *et al.*, 2010; Rosa *et al.*, 2009). Thus, it remains to be elucidated whether mechanisms exist to direct these miRNAs to alternate targets or there are other mechanisms to balance them as cells switch from a state that inhibits EMT (pluripotency) to a state that promotes EMT (mesendoderm differentiation).

B. miR-371-373 Cluster (miR-371, miR-372, miR-373)

The miR-371-373 cluster expresses a primary transcript in human ESCs that contains three stem-loops, including miR-371, miR-372, and miR-373. All transcripts are highly abundant in human ESCs and downregulated during undirected differentiation, thus displaying similar expression patterns to ESCC miRNAs described in mice. miR-371 has a unique seed sequence shifted by a single base, while miR-372 and miR-373 share the same seed sequence with the ESCC miRNAs described above. Although the ESCC miRNAs were experimentally proven to rescue cell proliferation in mouse ESCs specifically, the seed sequence is also the most abundantly expressed in human ESCs (Laurent *et al.*, 2008), and members of this cluster share some similar functions in maintenance of pluripotency. Human ESCs that are deficient in miRNA biogenesis exhibit decreased proliferation (Qi *et al.*, 2009). This defect can be partially rescued by ectopic miR-372 expression, which targets regulators of both the G1/S transition (CDKN1A) and the G2/M transition (WEE1) of the cell cycle (Qi *et al.*, 2009). Thus, miR-372 has similar biological roles to its homologous ESCC miRNAs in mouse ESCs.

miR-372 and miR-373 are homologous to members of the mouse miR-290–295 cluster that were identified as ESCC miRNAs and used to enhance reprogramming of fibroblasts to iPSCs. The role of miR-372 was investigated in enhancing human fibroblast reprogramming (Subramanyam *et al.*, 2011). As discussed above, both miR-302b and miR-372 were shown to target mRNAs involved in various pathways, including cell cycle regulation, epigenetic regulation, and EMT. Similar targets involved in cell cycle regulation and epigenetic regulation have also been implicated as targets by members of the homologous miR-290-295 family in mice. Further investigation into the MET pathway showed that ectopic expression of miR-372 during reprogramming resulted in decrease of the mesenchymal markers ZEB1, ZEB2, and SLUG, and subsequent increase in epithelial markers JAM-1 and E-cadherin (ECAD) (Subramanyam *et al.*, 2011). Studies by Zhou *et al.* (2011) demonstrated that β-Catenin/LEF1, members of the Wnt signaling pathway, directly activate transcription of the miR-371-373 cluster. miR-372 and miR-373 in turn target several genes implicated in the Wnt pathway (TGFBR2, BTG1) as well as the Wnt inhibitor DKK1 and Lefty, which was previously identified as a target for miR-302a (Rosa and

Brivanlou, 2009). Interestingly, members of the miR-371-373 cluster are specifically upregulated during directed differentiation to definitive endoderm (Hinton *et al.*, 2010), while Wnt and nodal/activin signaling pathways are important in both pluripotency and mesendoderm formation. Thus, similar to miR-302, members of this cluster are implicated in pathways that promote both EMT and MET.

C. miR-290-295 Cluster

The miR-290-295 cluster, which is expressed in mice but not in humans, consists of at least 14 mature miRNAs. Six of these miRNAs (miR-290-3p, miR-291a-3p, miR-291b-3p, miR-292-3p, miR-294, and miR-295) share the same seed sequence with other ESCC miRNAs described (Wang *et al.*, 2008), which were shown to rescue the ES cell cycle defect in DGCR8-null mouse ESC. Early miRNA expression profiling studies reported that the miR-290-295 cluster constitutes the majority of expressed miRNAs in mouse ESCs (Houbaviy *et al.*, 2003; Marson *et al.*, 2008), and in a similar fashion to the miR-302 cluster, these miRNAs were downregulated during undirected differentiation. Chromatin binding studies in mouse ESCs showed that SOX2, OCT4, TCF3, and Nanog co-occupy the promoter of the miR-290-295 cluster (Marson *et al.*, 2008), indicating positive regulation by the core pluripotency factors. The abundance of miRNAs with common seed sequences indicates that expression is coordinated to fine tune the regulation of similar targets.

An example of matched regulation of ESCC comes from studies of the regulated process of Cyclin E/Cdk2 during G1/S transition in mouse ESC. Wang *et al.* found that miRNAs function to derepress the Cyclin E/Cdk2 complex. Specifically, inhibitors of Cyclin E–CDK2 (p21, RBL2, and LATS2) were identified as direct targets of miR-291a-3p, miR-291b-3p, miR-294, miR-295, and miR-302 (Wang *et al.*, 2008). *Dcgr8* knockout cells consistently had higher levels of *Cdkn1a* (*p21*) and an increased G1 fraction. Upon stable transfection with p21, mouse ESCs displayed an increased G1 fraction very similar to that observed in *Dgcr8* knockout cells, consistent with *p21* as an important functional target of the miR-290-295 cluster. A separate study implicated *Wee1* as a target of the miR-290-295 cluster (Lichner *et al.*, 2011), implicating a role in regulation of the G2-M transition similar to that found with miR-372 in human ESCs (Qi *et al.*, 2009).

Members of the miR-290-295 cluster also play a role in epigenetic regulation, promoting the expression of *de novo* methylases (DNMT3s) in mouse ESCs via direct repression of Rbl2 (Benetti *et al.*, 2008). DNMT3s must be ready to silence specific genes rapidly upon differentiation both for lineage specification and for complete silencing of pluripotency factors, such as OCT4. Incomplete silencing of OCT4 is likely to be responsible for differentiation defects observed in DGCR8-null mouse ESCs (Wang *et al.*, 2007).

The miR-290-295 cluster indirectly activates self-renewal genes, *MYC* and *Lin28*, and was shown to improve reprogramming of somatic cells to iPSCs with methods involving these factors (Judson *et al.*, 2009; Melton *et al.*, 2010). miR-294 specifically enhances the efficiency of iPSC formation during reprogramming with OCT4, SOX2, and KLF4 but had no enhancement effect in the presence of c-Myc, indicating that ectopic miR-294 can substitute for c-Myc expression in somatic cell reprogramming (Judson *et al.*, 2009).

Members of this miRNA cluster that do not share the canonical seed also have roles in maintenance of pluripotency. miR-291b-5p and miR-293 directly repress p65, a subunit of nuclear factor kappa-light-chain-enhancer of activated B cells (NF-κB) (Luningschror *et al.*, 2012). Overexpression of p65 was shown to promote differentiation of mouse ESC as evidenced by expression markers of EMT (Luningschror *et al.*, 2012).

Zovoilis *et al.* (2009) showed that ectopic expression of miR-290-295 could directly inhibit DKK-1, an inhibitor of the Wnt signaling pathway. Although Wnt signaling promotes pluripotency, the study showed that overexpression of the miR-290-295 did not prevent differentiation but specifically prevented mesoderm formation in differentiating mouse ESCs, which is in accordance with the DGCR8 knockout study (Wang *et al.*, 2007). The inhibition of DKK-1 is similar to that described for homologous miR-372 in human ESCs (Zhou *et al.*, 2011). The nodal antagonist lefty was also identified as a target of the miR-290-295 cluster in mouse ESCs (Zovoilis *et al.*, 2009), similar to miR-302a in human ESCs (Rosa and Brivanlou, 2009). Similar to miR-302b and miR-372, transfection of mouse miR-294 mimic into human cells also inhibited TGF-β–induced EMT (Subramanyam *et al.*, 2011). The accumulation of these studies would indicate that ESCC miRNAs play a similar role in mouse and human ESCs for fine tuning the balance of Wnt and TGF-β/nodal signaling as the shift occurs between stem cell renewal and lineage specification.

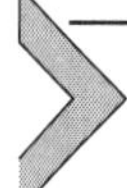

IV. miRNAs that Regulate Early Cell Fate Decisions in human ESC

Since their discovery, miRNAs have been shown to control developmental cell fate decisions. lin-4 and let-7, the first miRNAs to be discovered, are conserved across many species and control differentiation of hypodermal seam cell lineages in *Caenorhabditis elegans*. In animals lacking these miRNAs, the seam cells continue to maintain their stem cell–like fates and fail to terminally differentiate (Lee *et al.*, 1993; Reinhart *et al.*, 2000). Since the initial discovery, lin-4 and let-7 homologues, as well as other miRNAs, have been shown to serve key roles in differentiation and regulation of pluripotency. In this section, we will review the roles that mammalian let-7, miR-125 (*lin-4* homolog), miR-9, miR-145, and miR-34 families play in promoting ESC differentiation and influencing early decisions about cell fate. A diagram of how these miRNAs regulate their targets is shown in Fig. 1.2.

A. let-7

Soon after confirmation that let-7 controls *Caenorhabditis elegans* differentiation, it was also shown to be widely expressed in adult human tissues (Pasquinelli *et al.*, 2000). Other laboratories demonstrated that this miRNA had a conserved temporal expression pattern in mammals that increased as differentiation progressed across a wide variety of tissue types. Consistent with a critical role in cell differentiation, let-7 is frequently mutated in cancers and

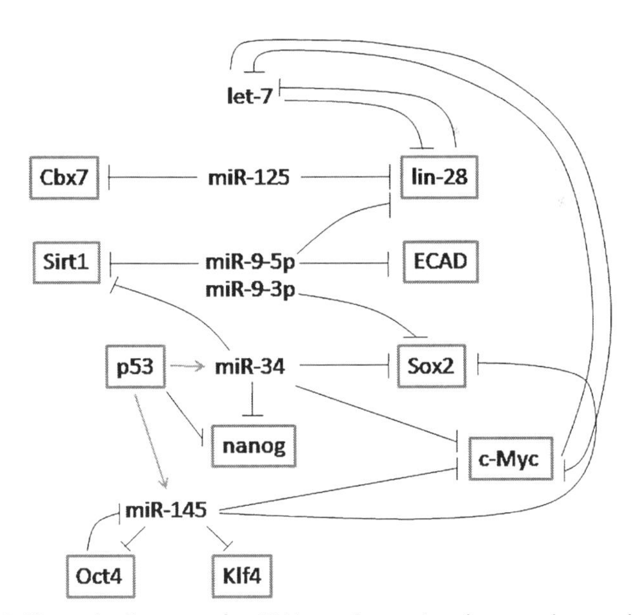

Figure 1.2 Network diagram of miRNAs and proteins that regulate early cell fate decisions in hESC. For color version of this figure, the reader is referred to the online version of this book.

its loss correlates with poor prognosis, especially in lung cancers (Inamura *et al.*, 2007; Jay *et al.*, 2007; Lu *et al.*, 2012; Raponi *et al.*, 2009; Takamizawa *et al.*, 2004). The targets of let-7 are key regulatory genes that control cell proliferation and differentiation, including HMG2A and members of the Myc and Ras families, suggesting that let-7 is important for preventing dedifferentiation necessary for tumor formation and progression (Christensen *et al.*, 2009; He *et al.*, 2010; Pan *et al.*, 2011; Sampson *et al.*, 2007).

Evidence supporting a role for let-7 in human ESC differentiation was demonstrated by ectopical expression of let-7 family members in mouse ESCs lacking DGCR8. In the absence of global miRNA maturation, specific expression of let-7 decreased expression of pluripotency markers and reduced the ability of the cells to reform ESC-like colonies (Melton and Blelloch, 2010). Introduction of ESCCs into DGCR8-null ESCs blocked the ability of let-7 to suppress self-renewal. Moreover, inhibition of let-7 enhanced somatic cell reprogramming, further supporting let-7 as a key regulator of differentiation. As such, there are multiple targets of let-7 that maintain pluripotency, including mutually inhibitory regulation with the pluripotency genes and somatic reprogramming factors Lin28 and c-Myc (Takahashi *et al.*, 2007; Yu *et al.*, 2011). Lin28 binds to the let-7 stem-loop to block production of mature let-7, whereas c-Myc both directly drives expression of Lin28 and represses the expression of the let-7a/d/f cluster (Chang *et al.*, 2009; Wang *et al.*, 2011), thus preventing let-7-mediated differentiation.

During differentiation, OCT4, SOX2, c-Myc, and NANOG levels decrease, leading to a downregulation of Lin28 expression, thus allowing let-7 biogenesis. Lin28 is also a let-7 target gene; therefore, as let-7 is produced, it binds and reduces the expression of Lin28 driving a positive feedback loop further reducing Lin28 levels. In addition to Lin28, let-7 directly targets c-Myc (Koscianska *et al.*, 2007; Sampson *et al.*, 2007), which simultaneously relieves direct transcriptional control by c-Myc- and lower c-Myc-driven Lin28 expression. Thus, let-7 disrupts the pluripotency maintenance network at multiple nodes.

B. miR-125

miR-125 is the *Caenorhabditis elegans* lin-4 homolog, and like let-7, plays a conserved role in development (Boissart *et al.*, 2012; Schulman *et al.*, 2005). In human ESCs, miR-125 expression is upregulated during differentiation, and ectopic expression of miR-125b in ESCs leads to loss of expression of pluripotency markers combined with a concomitant upregulation of differentiation markers, including SOX3 and Nkx2.2 (O'Loghlen *et al.*, 2012). Control of differentiation is carried out in part through the conserved direct targeting of *lin-28*/Lin28 by *lin-4*/miR-125 in both worms and mammals (Rybak *et al.*, 2008; Zhong *et al.*, 2010). miR-125 further promotes differentiation by targeting the Drosophila polycomb homolog Cbx7. As part of the polycomb repressive complex 1 (PRC1), Cbx7 reinforces the repressive H3K27me3 chromatin mark. Cbx7 is specifically expressed in ESCs in contrast to other orthologs (Cbx2, 4, 6, and 8) that come on during differentiation. Knockdown of Cbx7 levels leads to loss of pluripotency and upregulation of polycomb group target genes, including markers of differentiation.

miR-125 has a reciprocal negative regulatory relationship with the TGF-β signaling pathway. Inhibition of Activin and BMP signaling together induces miR-125 expression in differentiating human ESCs, while miR-125 targets SMAD4, a key transducer of both Activin and BMP signaling (Boissart *et al.*, 2012). SMAD4 plays a key role in self-renewal and resistance to differentiation signals (Avery *et al.*, 2010).

C. Other miRs that Regulate Differentiation

1. miR-9

miR-9 is upregulated during differentiation of human ESCs and is expressed during the formation of cells of both neuronal and pancreatic lineages (Joglekar *et al.*, 2009; Krichevsky *et al.*, 2006), suggesting a general role for miR-9 in differentiation. miR-9-5p and miR-9-3p directly target several genes controlling pluripotency in embryonic stems cells. Among the known miR-9-5p targets is ECAD, a marker for embryonic stem cells (Liu *et al.*, 2012; Ma *et al.*, 2010). ECAD has also been demonstrated to be an important regulator of pluripotency and can replace OCT4 during somatic cell reprogramming (Redmer *et al.*, 2011). In addition to ECAD, miR-9-5p also directly regulates Lin28 (Zhong *et al.*, 2010). The early expression of miR-9 in human ESC differentiation, from Day 4 in embryoid body differentiation, combined with direct regulation of Lin28, would position this miRNA to act upstream of let-7 to direct loss of pluripotency and induce differentiation. Consistent with this role, miR-9-5p has also been shown to target other genes that regulate pluripotency and differentiation in ESCs, including Sirt1 and REST (Packer *et al.*, 2008; Saunders *et al.*, 2010). Sirt1 is an HDAC that is downregulated during ESC differentiation. In undifferentiated ESCs, Sirt1 directly represses developmental gene expression (Calvanese *et al.*, 2010). Sirt1 can also deacetylate p53 in ESCs, which prevents its ability to translocate to the nucleus and promote differentiation, in part, through the down-regulation of Nanog (Calvanese *et al.*, 2010; Han *et al.*, 2008). The transcriptional repressor REST, along with cofactors including CoREST, is expressed in ESCs and differentiating neuronal precursors, where it binds to and represses developmental genes. Loss of REST in ESCs destabilizes the pluripotent state through the deregulation of these developmental genes (Soldati *et al.*, 2012). Thus, miR-9 regulation of REST would sensitize ESCs to differentiation cues, especially to neuronal lineages. The miRNA arising from the alternative arm of the hairpin during miRNA biogenesis, miR-9-3p, also has been shown to target CoREST and SOX2 (Jeon *et al.*, 2011; Packer *et al.*, 2008).

2. *miR-34*

Another important driver of terminal differentiation and cell cycle exit is miR-34. The role for this miRNA in opposing somatic reprogramming by p53 is highlighted by the increased iPSC reprogramming efficiency upon deletion of the miR-34 family (Choi *et al.*, 2011). The authors went on to show that miR-34 directly targets pluripotency/reprogramming factors Sox2, Nanog, and n-Myc. Additional information about the link between p53, miR-34, and regulation of pluripotency was recently reported by Kim *et al.* (2011c) who provided insight into the role the p53/miR-34 network plays in restraining canonical Wnt signaling cascades during organism development. Addition of Wnt3a is required for efficient differentiation of human ESC into definitive endoderm and less efficient formation of pancreatic precursors and endocrine cells is observed in human ESC that are not supplemented with this factor (D'Amour *et al.*, 2006). p53 represses Wnt activity through transactivation of miR-34. Targets of miR-34 in the Wnt/β-catenin pathway include Wnt1, Wnt3, LRP6, β-catenin, and LEF1. Loss of p53 function results in an increase of Wnt signaling mediated primarily by a loss of miR-34 upregulation. The miR-34 family has also been shown to regulate c-Myc and Sirt1 (Aranha *et al.*, 2011; Cannell and Bushell, 2010; Choi *et al.*, 2011; Siemens *et al.*, 2011; Yamakuchi and Lowenstein, 2009; Yamakuchi *et al.*, 2008).

3. *miR-145*

Like miR-34 and let-7, miR-145 is a tumor suppressor miRNA controlled by p53 (Sachdeva *et al.*, 2009; Suzuki *et al.*, 2009). Expression of miR-145 in human ESCs is both necessary and sufficient to repress pluripotency in differentiating human ESCs. miR-145-5p targets OCT4, SOX2, KLF4, and c-Myc to promote differentiation (Sachdeva *et al.*, 2009; Xu *et al.*, 2009). Moreover, the miR-145 promoter contains an OCT4 binding site and OCT4 represses reporter constructs containing this site, demonstrating reciprocal negative regulation between miRNA and target, similar to c-Myc and let-7.

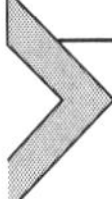

V. miRNAs THAT REGULATE ENDODERM AND PANCREATIC PRECURSOR FORMATION

A. miR-375

To date, miR-375 is the best-characterized miRNA in the development and function of cells of the pancreatic endoderm and β-cell. It is one of the few characterized tissue-specific miRNAs, and controls glucose-induced insulin secretion (Kloosterman *et al.*, 2007). In this section, we will focus on the role of this miRNA in early in endoderm development. Its function in β-cell genesis and insulin secretion are described later.

1. miR-375 functions in mesendoderm and definitive endoderm

Mesendoderm formation is one of the earliest events in gastrulation. It begins with the formation of the primitive streak in which epiblast cells mobilize and egress to later form separate endoderm and mesoderm layers (Tada *et al.*, 2005). This means that signals driving specification and lineage commitment are initially the same for these two cell populations, and the challenge remains to understand what drives specification of endoderm and mesoderm from mesendoderm progenitors. The definitive endoderm later gives rise to pancreas progenitors.

The importance of miR-375 in endocrine cell development became evident in studies with morpholino inhibitors in the zebrafish embryo (Kloosterman *et al.*, 2007). Injection of anti-miR-375 morpholinos into one- or two-stage embryos resulted in a disrupted islet cell phenotype and delayed development of endoderm derived exocrine pancreas, liver, and gut. A mouse study profiling miRNA expression in developing pancreas showed that miR-375 is expressed in the pancreatic ductal epithelium from which endocrine and exocrine cells arise. Furthermore, it colocalizes with PDX-1, a marker of pancreatic progenitors cells (Lynn *et al.*, 2007). In the same study, conditional deletion of the miRNA-processing enzyme Dicer1 in the pancreas resulted in a dramatic reduction in endocrine cells, especially of β-cells. Although in this system, all miRNAs are deleted from the developing pancreas, miR-375 was the most overrepresented miRNA in these tissues.

In human ESC, we and other investigators have tried to determine the specific signature of miRNAs at each stage of endodermal differentiation using microarrays and high-throughput sequencing. In these studies, miR-375 was consistently found to be highly expressed. Other highly expressed miRNAs included miR-122, members of the miR-302/367 cluster, and the miR-371/372/373 cluster (Hinton *et al.*, 2010; Tzur *et al.*, 2008). The functions of miR-375 in early endoderm formation and their overlap with the functions described in adult islets remain unknown. Furthermore, the targets of miR-375 in these progenitor cells remain a mystery. Only TIMM8A was found to be a putative target of miR-375, but its significance is unclear since it is has not been identified as a protein involved in signaling or developmental pathways (Hinton *et al.*, 2010). During human ESC differentiation, expression of miR-375 was highly elevated in mesendoderm-like stage and expressed even higher in definitive endoderm, but was not expressed in ectoderm or extraembryonic tissues, implicating a specific role for miR-375 in definitive endoderm specification (Laurent, 2008; Hinton *et al.*, 2010).

Two other signaling molecules implicated in cell development have recently been identified as targets of miR-375, yes-associated protein (YAP) and JAK2 (Ding *et al.*, 2010; Kowalik *et al.*, 2011; Liu *et al.*, 2010). YAP is an oncogene and critical mediator of the evolutionarily conserved Hippo signaling pathway that regulates proliferation and organ size through sensing changes in cell density. At low cell density, YAP induces expression of genes that result in cell growth and proliferation (Zhao *et al.*, 2011). In hepatic cancer cells, ectopic expression of miR-375 decreased endogenous YAP levels and diminished expression and blocked metastasis. Furthermore, miR-375 levels were significantly downregulated in patients with hepatic tumors compared with adjacent nontumor tissues of patients (Liu *et al.*, 2010). JAK2 regulates mouse ESC pluripotency through the LIF/gp130 signaling pathway. We have recently uncovered a similar signaling pathway in human ESC that actively maintains pluripotency. Results from Ding *et al.* (2010) demonstrate that miR-375 targets JAK2 in human gastric cancers and suggest that the dramatic increase in miR-375 levels observed during the transition of human ESC from pluripotency to definitive endoderm may regulate JAK2 levels. We have confirmed this to be the case (King *et al.*, unpublished results), suggesting that miR-375 acts on selected targets at specific stages of development.

miR-375 expression has also been detected during the endoderm stage in hepatocyte differentiation protocols (Kim *et al.*, 2011b). Levels of miR-375 remain elevated well into the hepatocyte lineage, but no targets were identified. Subsequently, miR-375 was demonstrated to bind the 3' UTR (untranslated region) and downregulate astrocyte elevated gene-1 (AEG-1) in hepatocellular carcinoma. Both RNAi-mediated knockdown of AEG-1 and miR-375 overexpression in hepatocellular carcinoma cells, suppressed tumor properties (He *et al.*, 2011). Altogether, these experiments revealed how essential miR-375 is during pancreatic development.

B. Other miRNAs Important for Formation of Mesendoderm and Definitive Endoderm

The miR-430/427/302 family, which is conserved across species, is known to regulate Nodal signaling. Nodal is essential for establishment of discrete cell fates among the three primary embryonic germ layers during left–right axis formation. Depending on the species, members of this miRNA family have distinct functions that lead to a similar consequence: a precise control of germ layer specification. In zebrafish, miR-430 targets both Nodal and its antagonist Lefties, thus modulating a gradient of morphogenic signaling (Choi *et al.*, 2007). In *Xenopus*, miR-430 depletion leads to severe mesodermal defects. During human ESC differentiation, miR-302a is upregulated in mesendoderm and definitive endoderm, but not in ectoderm or extraembryonic tissues (Hinton *et al.*, 2010). Overexpression of miR-302a in human ECSs during nondirected differentiation promotes the mesendodermal lineage at the expense of neuroectoderm formation by targeting lefties specifically (Rosa and Brivanlou, 2009).

The miR-200 family (miR-200a, miR-200b, miR-200c, miR-429, and miR-141) has been shown to influence EMT by targeting the transcription factors Smad interacting protein-1 (SIP1)/ZEB2 and ZEB1 in mesenchymal cells and breast cancer cell lines (Chng *et al.*, 2010; Gregory *et al.*, 2008a; Korpal and Kang, 2008). In the context of differentiating stem cells, SIP1 promotes neurectoderm formation at the expense of mesendoderm by inhibiting Smad2/3 mediation of Activin/Nodal signaling (Chng *et al.*, 2010). The miR-200 family was also implicated in a mouse ESC differentiation model in which the chromatin modifier nitric oxide (NO) is used to induce mesoderm formation (Rosati *et al.*, 2011). In this model, NO treatment caused induction of miR-200 family expression. Furthermore, ectopic expression of miR-200 family miRNAs, or knockdown of their targets SIP1 and ZEB1, recapitulated NO treatment (Rosati *et al.*, 2011).

miR-15 and miR-16 are regulators of early embryonic patterning. In *Xenopus*, they act as molecules that translate the early asymmetry in Wnt/β-catenin into the generation of a gradient of Nodal responsiveness by partitioning the Nodal receptor Acvr2a along the dorsoventral axis (Martello *et al.*, 2007). The expression of miR-15 family members is apparently also under the negative control of Wnt/β-catenin signaling. However, these miRNAs are not highly expressed in definitive endoderm derived from human ESCs (Hinton, 2010).

Deletion of miR-17 family members, miR-92, miR-106a, and miR-106b, also revealed developmental defects (Ventura *et al.*, 2008). Moreover, several members of the miR-17 family, including miR-17-5p, miR-20a, miR-93, and miR-106a, are specifically expressed in undifferentiated or differentiating ESCs (Houbaviy *et al.*, 2003; Suh *et al.*, 2004). miR-17-5p and miR-93 showed specific upregulation in differentiating mesoderm cells, where (at least) miR-93 suppresses STAT3 expression (Foshay and Gallicano, 2009).

In addition to regulation of pluripotency, miR-125 also regulates early lineage specification through the promotion of neural fates over mesendodermal lineages. Inactivation of Lin28 in ESCs by miR-125 or siRNAs leads to neural fate bias during ESC differentiation (Wang *et al.*, 2012). Inactivation of TGF-β signaling through SMAD4 by miR-125 further promotes loss of pluripotency and biases cells toward neural fates (Boissart *et al.*, 2012). Thus, miR-125 serves as a key regulator of development by coupling exit from pluripotency and lineage specification.

C. miRNAs Important for Definitive Endoderm Differentiation

Both miR-375 and miR-122 are highly expressed during endoderm formation. Therefore, the functions of other signals or miRNAs are necessary for further specification. Such signals include factors, Activin A and Wnt3a are important for induction to definitive endoderm (Kroon *et al.*, 2008). TGF-β, bone morphogenetic protein 4, epidermal growth factor, and vascular endothelial growth factor, which are secreted by endoderm and are proved to be important for the mesoderm induction in

ESC models of cardiogenesis (Behfar and Terzic, 2007). Temporal expression and suppression of these miRNAs are essential to determine lineage specification.

miR-125b/Lin28 axis is an important regulator of early lineage specification and cardiomyocyte differentiation of mouse ESCs (Wang *et al.*, 2012). Overexpression of miR-125b results in a reduction in mesodermal expression markers, inhibition of adipogenic and chondrogenic lineages, but not ectoderm differentiation. In mice, its direct target is Lin28, and downregulation of Lin28 is essential to maintain the undifferentiated state. During differentiation, the levels of miR-125b are decreased, and Lin28 is expressed to allow the mouse ESCs to differentiate to mesendoderm and further on into cardiomyocytes. Thus, the sharp decrease in miR-125b levels is required for specific lineage commitment.

VI. miRNAs THAT REGULATE PANCREATIC ENDOCRINE CELL FUNCTION

There is growing evidence that miRNAs function in all cell types to maintain homeostasis, regulate responses to external stimuli required for normal growth and differentiation as well as responses to stress and injury. In this section, we will review data supporting a role for miRNAs in both normal islet function and EMT. EMT in islets occurs after *in vitro* expansion of β-cells and results in loss of insulin production, creating a significant hurdle to supply these cells for therapeutic use. Therefore, the role of miRNAs in EMT is of particular interest to islet biologists because it provides a potential avenue to regulate insulin expression after expansion.

A. miR-375

miR-375 has been implicated not only in early human ESC development into cells of endocrine lineage (Hinton *et al.*, 2010) but also in mature islet function. Initially identified in the murine insulinoma β-cell line MIN-6, miR-375 was identified as an abundant pancreas-specific miRNA that regulates insulin secretion and maintains α- and β-cell mass (Poy *et al.*, 2004, 2009). Introduction of miR-375 into MIN-6 cells decreased their ability to secrete insulin in response to elevated glucose levels in a dose-dependent manner (Poy *et al.*, 2004). These studies were followed by analysis of pancreas development in miR-375 knockout mice (Poy *et al.*, 2009).

Ablation of miR-375 resulted in sustained hyperglycemia and was attributed to a significant increase in a cell mass and elevated glucagon release. Overexpression inhibited insulin secretion, and this effect was mimicked by knock down of its target *myotrophin*. Importantly, depletion of miR-375 increased the levels of its target and enhances glucose-induced insulin secretion (Poy *et al.*, 2004). Subsequently, miR-375 was identified as a regulator of pancreas development in zebra fish (Kloosterman *et al.*, 2007). A more thorough review of miR-375 function in the pancreas can be found in Baroukh and Van Obberghen (2009).

In humans, much less is known about the role of miR-375 in pancreas development and function. However, the role of this miRNA has been well documented in a range of human cancers, including gastric, esophageal, and small-cell lung cancers (Ding *et al.*, 2010; Li *et al.*, 2011; Zhao *et al.*, 2012). Targets identified from these studies, when combined with reports from mouse and other animal studies, allow us to speculate as to the role of miR-375 in the adult pancreas. The best-characterized target of miR-375 is 3-phosphoinositide-dependent kinase 1 (PDK-1), an important mediator of insulin signal transduction, that phosphorylates and activates protein kinases critical for cell growth, proliferation, movement, and survival (Toker and Newton, 2000). Because PDK-1 is constitutively active, its expression, localization, and access to substrates must be tightly controlled (Anderson *et al.*, 1998; Egawa *et al.*, 2002; King and Newton, 2004). Overexpression of miR-375 in the mouse INS-1E β-cells resulted in a 40% decrease in PDK-1 expression, decrease in AKT and GSK3β phosphorylation, and decrease in insulin synthesis (El Ouaamari *et al.*, 2008). Subsequently, Tsukamoto *et al.* (2010) surveyed changes in miRNA expression levels in human control and gastric cancer patients and identified PDK-1 as a target of miR-375. Taken together, these results suggest that PDK-1 is a conserved target of miR-375 in mouse and human. Specific knockout of PDK-1 in mouse β-cells resulted in hyperglycemia, loss of islet mass, and the development of overt diabetes (Hashimoto *et al.*, 2006). However, caution should be employed when translating mouse studies to human systems. Although miR-375-mediated changes in PDK-1 expression have been observed in both murine systems and primary human cancers, no changes in PDK-1 expression were observed in human ESC at multiple stages of development toward endocrine lineage after infection with miR-375 lentiviral constructs, suggesting that in humans, other miRNAs or regulatory factors exist during normal development (King *et al.*, unpublished results).

B. miR-7

miR-7 is the most abundant endocrine miRNA in rat and human islets (Bravo-Egana *et al.*, 2008). Levels in both species were found to be far greater than even miR-375, which is primarily restricted to β-cells (Joglekar *et al.*, 2009). Temporal expression of miR-7 in human fetal pancreas of gestational ages 8–22 weeks found a dramatic increase at weeks 14–18 that correspond to induction of *PDX-1, ISL-1,* and other genes required for endocrine cells fate specification (Correa-Medina *et al.*, 2009). These results are consistent with data from zebra fish that restrict miR-7 expression to endocrine cells (Wienholds *et al.*, 2005). Recently, the role of miR-7 was explored using antisense morpholinos delivered to an e10.5 day mouse embryo. This resulted in a downregulation of insulin production, decreased b-cells, and glucose intolerance after birth. However, to date, no targets of miR-7 have been identified.

C. miRNAs that Regulate Insulin Biosynthesis

Many recent studies have focused on the role of miRNAs in the regulation of insulin biosynthesis. Insulin expression in adult cells is controlled by a regulatory network of proteins that act to activate and repress transcription of the insulin. The observation that miRNAs regulate this network both by directly acting on the insulin genes and on the repressors/activators of the system adds an additional level of complexity to insulin biosynthesis.

In a screen of MIN6 cells for miRNAs regulated by changes in glucose levels, Tang *et al.* (2009) identified over 100 different miRNAs whose expression changed, including miR-30d. Overexpression of this miRNA increased insulin gene expression, while inhibition abolished glucose-stimulated insulin expression. Although these data suggest a role for miR-30d in downregulation of unidentified transcriptional repressor of the insulin gene, no targets were identified. An elegant study by Melkman-Zehavi *et al.* (2011) demonstrated the significance of miRNAs in the function of mature murine β-cells through knockout of dicer1. The phenotypic effect of dicer1 ablation in β-cells was hyperglycemia, acute glucose intolerance, and, most significantly, decreased expression of the insulin gene. Quantitative PCR was employed to quantify the expression levels of transcription factors, previously reported to function as repressors of insulin synthesis, including SOX6 and Bhlhe22/Beta3. miR-24, miR-26, miR-182, miR-148, and miR-200 were identified as regulators of SOX6 and Bhlhe22/Beta3, both these proteins have previously been implicated as negative regulators of insulin biosynthesis. Sox6 interferes with Pdx-1 transcriptional activation (Iguchi *et al.*, 2005) and Bhlhe22/Beta3 (Peyton *et al.*, 1996). In human cells, miR-26a/b has been implicated in cell cycle control, but not in pancreatic development. Expression of miR-182 and miR-200 were identified in a screen of errantly expressed miRNAs in different pancreatic intraepithelial neoplasias compared with normal pancreatic duct samples, suggesting that these miRNAs may play a role in maintenance of normal islet function in humans (Yu *et al.*, 2012).

Although the miR17-92 family is primarily a regulator of stem cell pluripotency, a function of miR-19b, a member of the miR17-92 family, was recently identified in mouse and human cells (Zhang *et al.*, 2011). These studies utilized the PicTar and TargetScan algorithms to select NeuroD as a potential target based on seed sequence similarity. Expression of miR-19b was confirmed in both MIN6 and primary mouse islets and levels of miR-19b decreased during gestation. There was no effect of miR-19b on proliferation of MIN6 cells, but levels of insulin 1, not insulin 2, were significantly decreased upon overexpression of this miRNA.

D. miRNAs that Regulate EMT in Pancreatic Cells

Both human adult islets and fetal islet cell clusters can be successfully expanded when grown *in vitro* as monolayers on a matrix in the presence of the growth factor hepatocyte growth factor/scatter factor (Beattie *et al.*, 1997, 1999; Beattie *et al.*). Previous studies have demonstrated that both cell populations can be reaggregated; however, only human fetal islet cells will form functional islets when transplanted under the kidney capsule of nude mice (Beattie *et al.*, 1994; Hayek and Beattie, 1997). These results suggest that there is a fundamental difference in expansion of fetal islets versus expansion of human adult islets. The underlying biochemistry that regulates this process lies in the elementary difference between the two cell populations. Functional human adult islets exist as terminally differentiated epithelial cells. Upon expansion, the cells undergo a fundamental change from epithelial to mesenchymal cells, EMT. This is a developmental program, characterized by a loss of cell adhesion and increased cell mobility—changes required for cell proliferation. Subsequently, the expanded mesenchymal cells must revert back to epithelial cells, MET, before becoming functional insulin-secreting cells. This transition has been a roadblock for generating insulin-producing cells despite of claims from several laboratories that the process is reversible (Beattie *et al.*, 2002; Gershengorn *et al.*, 2004; Kayali *et al.*, 2007; Ouziel-Yahalom *et al.*, 2006). Therefore, the observation that miRNAs may regulate EMT provides additional potential targets to regulate this process. Studies have demonstrated a link between miRNAs, EMT, and cancer (Gregory *et al.*, 2008b), with a specific role for the miR-200 family (Gregory *et al.*, 2011; Paterson *et al.*, 2008), although there are no established links to normal pancreas development or function. However, the miR-30 family appears to regulate EMT in human fetal pancreatic cells (Joglekar *et al.*, 2009). In this study, endogenous levels of miR-30 increased during the transition of mesenchymal cells derived from islets into fetal islet cell clusters. Infection with antisense miRNAs to bind endogenous miR-30 enhanced the mesenchymal phenotype, while ectopic overexpression of miR-30 resulted in maintenance of the epithelial phenotype. The authors validated a number of known EMT targets of miR-30, including vimentin and SNAIL1.

A growing body of evidence implicates the miR-200 family in regulation of EMT. Initially identified as regulators of SIP1/ZEB1 expression, which act as transcriptional repressors of ECAD, the miR-200 family has recently been shown to target a number of additional genes that regulate EMT, including FN1 and moesin (Howe *et al.*, 2011). Loss of miR-200c was observed in breast cancer cells and, upon restoration, decreases were observed in nonepithelial, EMT-associated genes. These observations are in line with those of Chang *et al.* (2011) who demonstrated that p53 induces miR-200c to suppress EMT through inhibition of ZEB1 and decrease "stemness" through inhibition of BMI1. Interestingly, ZEB1 is able to reciprocally regulate miR-200c expression in, and therefore regulate, EMT. This feedback loop was recently demonstrated to be part of a larger feedback loop that is centered about p53 and involves a second miRNA, miR-34 (Siemens *et al.*, 2011). The crux of the problem was that miR-200 family members could regulate expression of ZEB1, but not the potent EMT transducer SNAIL. p53 activation induced MET through upregulation of miR-34, which directly targeted the mesenchymal factor SNAIL. Additionally, overexpression of SNAIL repressed miR-34 expression, connecting the feedback loop. A diagram of this feedback mechanism is shown in Fig. 1.3.

The implication from these findings provides a path to explore how islet expansion is regulated. The exquisite control of miR-34 and miR-200

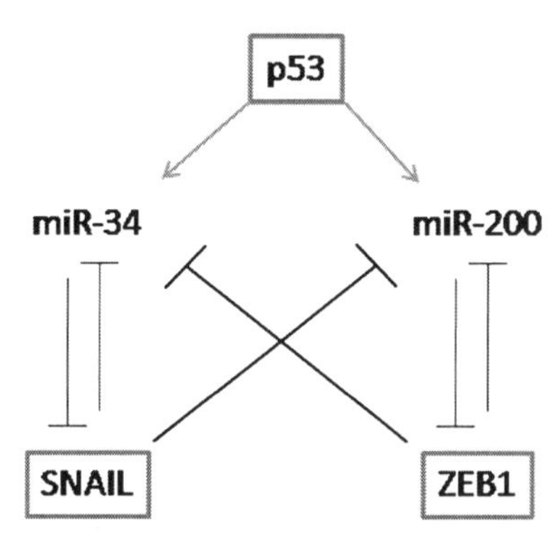

Figure 1.3 Network diagram of the interaction of p53 with miRNAs that regulate Zeb1 and SNAIL. For color version of this figure, the reader is referred to the online version of this book.

expression by p53 and downstream regulators of EMT suggests that biological intervention to ectopically tune expression of both miR-200 and miR-34 during EMT may help generate therapeutically β-cells.

VII. CONCLUSIONS AND FUTURE DIRECTIONS

Despite all of the above advances, the function of most miRNAs detected in stem cells remains poorly characterized. In this chapter, we have surveyed the increasing literature on the role of specific miRNAs during differentiation of hESC into pancreatic precursors, endocrine cells, and in human adult islets with the hope that connections between hubs of information will help further elucidate the role other miRNAs play in differentiation and cell function. Advances in deep sequencing continue to provide a wealth of additional information about the dynamic levels of miRNAs in differentiation; however, more information about the role of, isomiRs, snoRNAs, and other small noncoding RNAs is required. These efforts, along with more information about spatiotemporal dynamics of miRNA targeting, will ultimately determine our ability to use miRNAs as candidate therapeutics for diseases, such as diabetes.

One approach to develop a stem cell-based therapy for the treatment of type 1 diabetes, includes employing computational genomics and biochemical analyses to link changes in miRNA expression profiles of purified pancreatic progenitor and insulin-producing cells derived from hESCs to therapeutically relevant outputs, including *in vitro* cell differentiation, proliferation, and apoptosis, as well as function *in vivo* after experimental transplantation.

ACKNOWLEDGMENTS

The authors thank Alberto Hayek for critical input in the preparation of this manuscript. This work was supported by a grant to C.C.K. from the California Institute for Regenerative Medicine (CIRM).

REFERENCES

Anderson, K.E., Coadwell, J., Stephens, L.R., Hawkins, P.T., 1998. Translocation of PDK-1 to the plasma membrane is important in allowing PDK-1 to activate protein kinase B. Curr. Biol. 8 (12), 684–691.

Anokye-Danso, F., Trivedi, C.M., Juhr, D., Gupta, M., Cui, Z., Tian, Y., Zhang, Y., Yang, W., Gruber, P.J., Epstein, J.A., Morrisey, E.E., 2011. Highly efficient miRNA-mediated reprogramming of mouse and human somatic cells to pluripotency. Cell Stem Cell 8 (4), 376–388.

Aranha, M.M., Santos, D.M., Sola, S., Steer, C.J., Rodrigues, C.M., 2011. miR-34a regulates mouse neural stem cell differentiation. PLoS One 6 (8), e21396.

Avery, S., Zafarana, G., Gokhale, P.J., Andrews, P.W., 2010. The role of SMAD4 in human embryonic stem cell self-renewal and stem cell fate. Stem Cells 28 (5), 863–873.

Bagga, S., Pasquinelli, A.E., 2006. Identification and analysis of microRNAs. Genet. Eng. (N Y) 27, 1–20.

Bar, M., Wyman, S.K., Fritz, B.R., Qi, J., Garg, K.S., Parkin, R.K., Kroh, E.M., Bendoraite, A., Mitchell, P.S., Nelson, A.M., Ruzzo, W.L., Ware, C., Radich, J.P., Gentleman, R., Ruohola-Baker, H., Tewari, M., 2008. MicroRNA discovery and profiling in human embryonic stem cells by deep sequencing of small RNA libraries. Stem Cells 26 (10), 2496–2505.

Baroukh, N.N., Van Obberghen, E., 2009. Function of microRNA-375 and microRNA-124a in pancreas and brain. FEBS J. 276 (22), 6509–6521.

Barroso-del Jesus, A., Lucena-Aguilar, G., Menendez, P., 2009. The miR-302-367 cluster as a potential stemness regulator in ESCs. Cell Cycle 8 (3), 394–398.

Bartel, D.P., 2004. MicroRNAs: genomics, biogenesis, mechanism, and function. Cell 116 (2), 281–297.

Bartel, D.P., 2009. MicroRNAs: target recognition and regulatory functions. Cell 136 (2), 215–233.

Beattie, G.M., Butler, C., Hayek, A., 1994. Morphology and function of cultured human fetal pancreatic cells transplanted into athymic mice: a longitudinal study. Cell Transplant. 3 (5), 421–425.

Beattie, G.M., Cirulli, V., Lopez, A.D., Hayek, A., 1997. Ex vivo expansion of human pancreatic endocrine cells. J. Clin. Endocrinol. Metab. 82 (6), 1852–1856.

Beattie, G.M., Itkin-Ansari, P., Cirulli, V., Leibowitz, G., Lopez, A.D., Bossie, S., Mally, M.I., Levine, F., Hayek, A., 1999. Sustained proliferation of PDX-1+ cells derived from human islets. Diabetes 48 (5), 1013–1019.

Beattie, G.M., Lopez, A.D., Bucay, N., Hinton, A., Firpo, M.T., King, C.C., Hayek, A., 2005. Activin A maintains pluripotency of human embryonic stem cells in the absence of feeder layers. Stem Cells 23 (4), 489–495.

Beattie, G.M., Montgomery, A.M., Lopez, A.D., Hao, E., Perez, B., Just, M.L., Lakey, J.R., Hart, M.E., Hayek, A., 2002. A novel approach to increase human islet cell mass while preserving beta-cell function. Diabetes 51 (12), 3435–3439.

Beattie, G.M., Rubin, J.S., Mally, M.I., Otonkoski, T., Hayek, A., 1996. Regulation of proliferation and differentiation of human fetal pancreatic islet cells by extracellular matrix, hepatocyte growth factor, and cell-cell contact. Diabetes 45 (9), 1223–1228.

Behfar, A., Terzic, A., 2007. Cardioprotective repair through stem cell-based cardiopoiesis. J. Appl. Physiol. 103 (4), 1438–1440.

Benetti, R., Gonzalo, S., Jaco, I., Munoz, P., Gonzalez, S., Schoeftner, S., Murchison, E., Andl, T., Chen, T., Klatt, P., Li, E., Serrano, M., Millar, S., Hannon, G., Blasco, M.A., 2008. A mammalian microRNA cluster controls DNA methylation and telomere recombination via Rbl2-dependent regulation of DNA methyltransferases. Nat. Struct. Mol. Biol. 15 (9), 998.

Boissart, C., Nissan, X., Giraud-Triboult, K., Peschanski, M., Benchoua, A., 2012. miR-125 potentiates early neural specification of human embryonic stem cells. Development.

Bravo-Egana, V., Rosero, S., Molano, R.D., Pileggi, A., Ricordi, C., Dominguez-Bendala, J., Pastori, R.L., 2008. Quantitative differential expression analysis reveals miR-7 as major islet microRNA. Biochem. Biophys. Res. Commun. 366 (4), 922–926.

Calvanese, V., Lara, E., Suarez-Alvarez, B., Abu Dawud, R., Vazquez-Chantada, M., Martinez-Chantar, M.L., Embade, N., Lopez-Nieva, P., Horrillo, A., Hmadcha, A., Soria, B., Piazzolla, D., Herranz, D., Serrano, M., Mato, J.M., Andrews, P.W., Lopez-Larrea, C., Esteller, M., Fraga, M.F., 2010. Sirtuin 1 regulation of developmental genes during differentiation of stem cells. Proc. Natl. Acad. Sci. U. S. A. 107 (31), 13736–13741.

Cannell, I.G., Bushell, M., 2010. Regulation of Myc by miR-34c: a mechanism to prevent genomic instability? Cell Cycle 9 (14), 2726–2730.

Card, D.A., Hebbar, P.B., Li, L., Trotter, K.W., Komatsu, Y., Mishina, Y., Archer, T.K., 2008. Oct4/Sox2-regulated miR-302 targets cyclin D1 in human embryonic stem cells. Mol. Cell. Biol. 28 (20), 6426–6438.

Chang, C.-J., Chao, C.-H., Xia, W., Yang, J.-Y., Xiong, Y., Li, C.-W., Yu, W.-H., Rehman, S.K., Hsu, J.L., Lee, H.-H., Liu, M., Chen, C.-T., Yu, D., Hung, M.-C., 2011. p53 regulates epithelial-mesenchymal transition and stem cell properties through modulating miRNAs. Nat. Cell Biol. 13 (3), 317–323.

Chang, T.C., Zeitels, L.R., Hwang, H.W., Chivukula, R.R., Wentzel, E.A., Dews, M., Jung, J., Gao, P., Dang, C.V., Beer, M.A., Thomas-Tikhonenko, A., Mendell, J.T., 2009. Lin-28B transactivation is necessary for Myc-mediated let-7 repression and proliferation. Proc. Natl. Acad. Sci. U. S. A. 106 (9), 3384–3389.

Chng, Z., Teo, A., Pedersen, R.A., Vallier, L., 2010. SIP1 mediates cell-fate decisions between neuroectoderm and mesendoderm in human pluripotent stem cells. Cell Stem Cell 6 (1), 59–70.

Choi, W.-Y., Giraldez, A.J., Schier, A.F., 2007. Target protectors reveal dampening and balancing of nodal agonist and antagonist by miR-430. Science 318 (5848), 271–274.

Choi, Y.J., Lin, C.P., Ho, J.J., He, X., Okada, N., Bu, P., Zhong, Y., Kim, S.Y., Bennett, M.J., Chen, C., Ozturk, A., Hicks, G.G., Hannon, G.J., He, L., 2011. miR-34 miRNAs provide a barrier for somatic cell reprogramming. Nat. Cell Biol. 13 (11), 1353–1360.

Christensen, B.C., Moyer, B.J., Avissar, M., Ouellet, L.G., Plaza, S.L., McClean, M.D., Marsit, C.J., Kelsey, K.T., 2009. A let-7 microRNA-binding site polymorphism in the KRAS 3' UTR is associated with reduced survival in oral cancers. Carcinogenesis 30 (6), 1003–1007.

Correa-Medina, M., Bravo-Egana, V., Rosero, S., Ricordi, C., Edlund, H., Diez, J., Pastori, R.L., 2009. MicroRNA miR-7 is preferentially expressed in endocrine cells of the developing and adult human pancreas. Gene Expr. Patterns 9 (4), 193–199.

D'Amour, K.A., Agulnick, A.D., Eliazer, S., Kelly, O.G., Kroon, E., Baetge, E.E., 2005. Efficient differentiation of human embryonic stem cells to definitive endoderm. Nat. Biotechnol. 23 (12), 1534–1541.

Ding, L., Xu, Y., Zhang, W., Deng, Y., Si, M., Du, Y., Yao, H., Liu, X., Ke, Y., Si, J., Zhou, T., 2010. MiR-375 frequently downregulated in gastric cancer inhibits cell proliferation by targeting JAK2. Cell Res. 20 (7), 784–793.

Egawa, K., Maegawa, H., Shi, K., Nakamura, T., Obata, T., Yoshizaki, T., Morino, K., Shimizu, S., Nishio, Y., Suzuki, E., Kashiwagi, A., 2002. Membrane localization of 3-phosphoinositide-dependent protein kinase-1 stimulates activities of Akt and atypical protein kinase C but does not stimulate glucose transport and glycogen synthesis in 3T3-L1 adipocytes. J. Biol. Chem. 277 (41), 38863–38869.

El Ouaamari, A., Baroukh, N., Martens, G.A., Lebrun, P., Pipeleers, D., van Obberghen, E., 2008. miR-375 targets 3'-phosphoinositide-dependent protein kinase-1 and regulates glucose-induced biological responses in pancreatic beta-cells. Diabetes 57 (10), 2708–2717.

Foshay, K.M., Gallicano, G.I., 2009. miR-17 family miRNAs are expressed during early mammalian development and regulate stem cell differentiation. Dev. Biol. 326 (2), 431–443.

Gershengorn, M.C., Hardikar, A.A., Wei, C., Geras-Raaka, E., Marcus-Samuels, B., Raaka, B.M., 2004. Epithelial-to-mesenchymal transition generates proliferative human islet precursor cells. Science 306 (5705), 2261–2264.

Gregory, P.A., Bert, A.G., Paterson, E.L., Barry, S.C., Tsykin, A., Farshid, G., Vadas, M.A., Khew-Goodall, Y., Goodall, G.J., 2008a. The miR-200 family and miR-205 regulate epithelial to mesenchymal transition by targeting ZEB1 and SIP1. Nat. Cell Biol. 10 (5), 593–601.

Gregory, P.A., Bracken, C.P., Bert, A.G., Goodall, G.J., 2008b. MicroRNAs as regulators of epithelial-mesenchymal transition. Cell Cycle 7 (20), 3112–3118.

Gregory, P.A., Bracken, C.P., Smith, E., Bert, A.G., Wright, J.A., Roslan, S., Morris, M., Wyatt, L., Farshid, G., Lim, Y.Y., Lindeman, G.J., Shannon, M.F., Drew, P.A., Khew-Goodall, Y., Goodall, G.J., 2011. An autocrine TGF-beta/ZEB/miR-200 signaling network regulates establishment and maintenance of epithelial-mesenchymal transition. Mol. Biol. Cell 22 (10), 1686–1698.

Han, M.K., Song, E.K., Guo, Y., Ou, X., Mantel, C., Broxmeyer, H.E., 2008. SIRT1 regulates apoptosis and Nanog expression in mouse embryonic stem cells by controlling p53 subcellular localization. Cell Stem Cell 2 (3), 241–251.

Hashimoto, N., Kido, Y., Uchida, T., Asahara, S., Shigeyama, Y., Matsuda, T., Takeda, A., Tsuchihashi, D., Nishizawa, A., Ogawa, W., Fujimoto, Y., Okamura, H., Arden, K.C., Herrera, P.L., Noda, T., Kasuga, M., 2006. Ablation of PDK1 in pancreatic beta cells induces diabetes as a result of loss of beta cell mass. Nat. Genet. 38 (5), 589–593.

Hayek, A., Beattie, G.M., 1997. Experimental transplantation of human fetal and adult pancreatic islets. J. Clin. Endocrinol. Metab. 82 (8), 2471–2475.

He, X.X., Chang, Y., Meng, F.Y., Wang, M.Y., Xie, Q.H., Tang, F., Li, P.Y., Song, Y.H., Lin, J.S., 2011. MicroRNA-375 targets AEG-1 in hepatocellular carcinoma and suppresses liver cancer cell growth in vitro and in vivo. Oncogene.

He, X.Y., Chen, J.X., Zhang, Z., Li, C.L., Peng, Q.L., Peng, H.M., 2010. The let-7a microRNA protects from growth of lung carcinoma by suppression of k-Ras and c-Myc in nude mice. J. Cancer Res. Clin. Oncol. 136 (7), 1023–1028.

Hinton, A., Afrikanova, I., Wilson, M., King, C.C., Maurer, B., Yeo, G.W., Hayek, A., Pasquinelli, A.E., 2010. A distinct microRNA signature for definitive endoderm derived from human embryonic stem cells. Stem Cells Dev. 19 (6), 797–807.

Hohjoh, H., Fukushima, T., 2007. Marked change in microRNA expression during neuronal differentiation of human teratocarcinoma NTera2D1 and mouse embryonal carcinoma P19 cells. Biochem. Biophys. Res. Commun. 362 (2), 360–367.

Houbaviy, H.B., Murray, M.F., Sharp, P.A., 2003. Embryonic stem cell-specific microRNAs. Dev. Cell 5 (2), 351–358.

Howe, E.N., Cochrane, D.R., Richer, J.K., 2011. Targets of miR-200c mediate suppression of cell motility and anoikis resistance. Breast Cancer Res. 13 (2), R45.

Iguchi, H., Ikeda, Y., Okamura, M., Tanaka, T., Urashima, Y., Ohguchi, H., Takayasu, S., Kojima, N., Iwasaki, S., Ohashi, R., Jiang, S., Hasegawa, G., Ioka, R.X., Magoori, K., Sumi, K., Maejima, T., Uchida, A., Naito, M., Osborne, T.F., Yanagisawa, M., Yamamoto, T.T., Kodama, T., Sakai, J., 2005. SOX6 attenuates glucose-stimulated insulin secretion by repressing PDX1 transcriptional activity and is down-regulated in hyperinsulinemic obese mice. J. Biol. Chem. 280 (45), 37669–37680.

Inamura, K., Togashi, Y., Nomura, K., Ninomiya, H., Hiramatsu, M., Satoh, Y., Okumura, S., Nakagawa, K., Ishikawa, Y., 2007. let-7 microRNA expression is reduced in bronchioloalveolar carcinoma, a non-invasive carcinoma, and is not correlated with prognosis. Lung Cancer 58 (3), 392–396.

Ivey, K.N., Muth, A., Arnold, J., King, F.W., Yeh, R.F., Fish, J.E., Hsiao, E.C., Schwartz, R.J., Conklin, B.R., Bernstein, H.S., Srivastava, D., 2008. MicroRNA regulation of cell lineages in mouse and human embryonic stem cells. Cell Stem Cell 2 (3), 219–229.

James, D., Levine, A.J., Besser, D., Hemmati-Brivanlou, A., 2005. TGFbeta/activin/nodal signaling is necessary for the maintenance of pluripotency in human embryonic stem cells. Development 132 (6), 1273–1282.

Jay, C., Nemunaitis, J., Chen, P., Fulgham, P., Tong, A.W., 2007. miRNA profiling for diagnosis and prognosis of human cancer. DNA Cell Biol. 26 (5), 293–300.

Jeon, H.M., Sohn, Y.W., Oh, S.Y., Kim, S.H., Beck, S., Kim, S., Kim, H., 2011. ID4 imparts chemoresistance and cancer stemness to glioma cells by derepressing miR-9*-mediated suppression of SOX2. Cancer Res. 71 (9), 3410–3421.

Joglekar, M.V., Joglekar, V.M., Hardikar, A.A., 2009. Expression of islet-specific microRNAs during human pancreatic development. Gene Expr. Patterns 9 (2), 109–113.

Judson, R.L., Babiarz, J.E., Venere, M., Blelloch, R., 2009. Embryonic stem cell-specific microRNAs promote induced pluripotency. Nat. Biotechnol. 27 (5), 459–461.

Kanellopoulou, C., Muljo, S.A., Kung, A.L., Ganesan, S., Drapkin, R., Jenuwein, T., Livingston, D.M., Rajewsky, K., 2005. Dicer-deficient mouse embryonic stem cells are defective in differentiation and centromeric silencing. Genes Dev. 19 (4), 489–501.

Kayali, A.G., Flores, L.E., Lopez, A.D., Kutlu, B., Baetge, E., Kitamura, R., Hao, E., Beattie, G.M., Hayek, A., 2007. Limited capacity of human adult islets expanded in vitro to redifferentiate into insulin-producing beta-cells. Diabetes 56 (3), 703–708.

Kim, H., Lee, G., Ganat, Y., Papapetrou, E.P., Lipchina, I., Socci, N.D., Sadelain, M., Studer, L., 2011a. miR-371-3 expression predicts neural differentiation propensity in human pluripotent stem cells. Cell Stem Cell 8 (6), 695–706.

Kim, N., Kim, H., Jung, I., Kim, Y., Kim, D., Han, Y.-M., 2011b. Expression profiles of miRNAs in human embryonic stem cells during hepatocyte differentiation. Hepatol. Res. 41 (2), 170–183.

Kim, N.H., Kim, H.S., Kim, N.G., Lee, I., Choi, H.S., Li, X.Y., Kang, S.E., Cha, S.Y., Ryu, J.K., Na, J.M., Park, C., Kim, K., Lee, S., Gumbiner, B.M., Yook, J.I., Weiss, S.J., 2011c. p53 and microRNA-34 are suppressors of canonical Wnt signaling. Sci. Signal. 4 (197), ra71.

King, C.C., Newton, A.C., 2004. The adaptor protein Grb14 regulates the localization of 3-phosphoinositide-dependent kinase-1. J. Biol. Chem. 279 (36), 37518–37527.

Kloosterman, W.P., Lagendijk, A.K., Ketting, R.F., Moulton, J.D., Plasterk, R.H., 2007. Targeted inhibition of miRNA maturation with morpholinos reveals a role for miR-375 in pancreatic islet development. PLoS Biol. 5 (8), e203.

Korpal, M., Kang, Y., 2008. The emerging role of miR-200 family of microRNAs in epithelial-mesenchymal transition and cancer metastasis. RNA Biol. 5 (3), 115–119.

Koscianska, E., Baev, V., Skreka, K., Oikonomaki, K., Rusinov, V., Tabler, M., Kalantidis, K., 2007. Prediction and preliminary validation of oncogene regulation by miRNAs. BMC Mol. Biol. 8, 79.

Kowalik, M.A., Saliba, C., Pibiri, M., Perra, A., Ledda-Columbano, G.M., Sarotto, I., Ghiso, E., Giordano, S., Columbano, A., 2011. Yes-associated protein regulation of adaptive liver enlargement and hepatocellular carcinoma development in mice. Hepatology 53 (6), 2086–2096.

Krichevsky, A.M., Sonntag, K.C., Isacson, O., Kosik, K.S., 2006. Specific microRNAs modulate embryonic stem cell-derived neurogenesis. Stem Cells 24 (4), 857–864.

Kroon, E., Martinson, L.A., Kadoya, K., Bang, A.G., Kelly, O.G., Eliazer, S., Young, H., Richardson, M., Smart, N.G., Cunningham, J., Agulnick, A.D., D'Amour, K.A., Carpenter, M.K., Baetge, E.E., 2008. Pancreatic endoderm derived from human embryonic stem cells generates glucose-responsive insulin-secreting cells in vivo. Nat. Biotechnol. 26 (4), 443–452.

Kubo, A., Shinozaki, K., Shannon, J.M., Kouskoff, V., Kennedy, M., Woo, S., Fehling, H.J., Keller, G., 2004. Development of definitive endoderm from embryonic stem cells in culture. Development 131 (7), 1651–1662.

Lakshmipathy, U., Love, B., Goff, L.A., Jornsten, R., Graichen, R., Hart, R.P., Chesnut, J.D., 2007. MicroRNA expression pattern of undifferentiated and differentiated human embryonic stem cells. Stem Cells Dev. 16 (6), 1003–1016.

Laurent, L.C., 2008. MicroRNAs in embryonic stem cells and early embryonic development. J. Cell. Mol. Med. 12 (6A), 2181–2188.

Laurent, L.C., Chen, J., Ulitsky, I., Mueller, F.J., Lu, C., Shamir, R., Fan, J.B., Loring, J.F., 2008. Comprehensive microRNA profiling reveals a unique human embryonic stem cell signature dominated by a single seed sequence. Stem Cells 26 (6), 1506–1516.

Lee, R.C., Feinbaum, R.L., Ambros, V., 1993. The C. elegans heterochronic gene lin-4 encodes small RNAs with antisense complementarity to lin-14. Cell 75 (5), 843–854.

Li, R., Liang, J., Ni, S., Zhou, T., Qing, X., Li, H., He, W., Chen, J., Li, F., Zhuang, Q., Qin, B., Xu, J., Li, W., Yang, J., Gan, Y., Qin, D., Feng, S., Song, H., Yang, D., Zhang, B., Zeng, L., Lai, L., Esteban, M.A., Pei, D., 2010. A mesenchymal-to-epithelial transition initiates and is required for the nuclear reprogramming of mouse fibroblasts. Cell Stem Cell 7 (1), 51–63.

Li, X., Lin, R., Li, J., 2011. Epigenetic silencing of microRNA-375 regulates PDK1 expression in esophageal cancer. Dig. Dis. Sci. 56 (10), 2849–2856.

Lichner, Z., Pall, E., Kerekes, A., Pallinger, E., Maraghechi, P., Bosze, Z., Gocza, E., 2011. The miR-290-295 cluster promotes pluripotency maintenance by regulating cell cycle phase distribution in mouse embryonic stem cells. Differentiation 81 (1), 11–24.

Lipchina, I., Elkabetz, Y., Hafner, M., Sheridan, R., Mihailovic, A., Tuschl, T., Sander, C., Studer, L., Betel, D., 2011. Genome-wide identification of microRNA targets in human ES cells reveals a role for miR-302 in modulating BMP response. Genes Dev. 25 (20), 2173–2186.

Liu, A.M., Poon, R.T., Luk, J.M., 2010. MicroRNA-375 targets Hippo-signaling effector YAP in liver cancer and inhibits tumor properties. Biochem. Biophys. Res. Commun. 394 (3), 623–627.

Liu, S., Kumar, S.M., Lu, H., Liu, A., Yang, R., Pushparajan, A., Guo, W., Xu, X., 2012. MicroRNA-9 up-regulates E-cadherin through inhibition of NF-kappaB1-Snail1 pathway in melanoma. J. Pathol. 226 (1), 61–72.

Lu, Y., Govindan, R., Wang, L., Liu, P.Y., Goodgame, B., Wen, W., Sezhiyan, A., Li, Y.F., Hua, X., Wang, Y., Yang, P., You, M., 2012. MicroRNA profiling and prediction of recurrence/relapse-free survival in stage I lung cancer. Carcinogenesis.

Luningschror, P., Stocker, B., Kaltschmidt, B., Kaltschmidt, C., 2012. miR-290 cluster modulates pluripotency by repressing canonical NF-kappaB signaling. Stem Cells.

Lynn, F.C., Skewes-Cox, P., Kosaka, Y., McManus, M.T., Harfe, B.D., German, M.S., 2007. MicroRNA expression is required for pancreatic islet cell genesis in the mouse. Diabetes 56 (12), 2938–2945.

Ma, L., Young, J., Prabhala, H., Pan, E., Mestdagh, P., Muth, D., Teruya-Feldstein, J., Reinhardt, F., Onder, T.T., Valastyan, S., Westermann, F., Speleman, F., Vandesompele, J., Weinberg, R.A., 2010. miR-9, a MYC/MYCN-activated microRNA, regulates E-cadherin and cancer metastasis. Nat. Cell Biol. 12 (3), 247–256.

Mallanna, S.K., Rizzino, A., 2010. Emerging roles of microRNAs in the control of embryonic stem cells and the generation of induced pluripotent stem cells. Dev. Biol. 344 (1), 16–25.

Marson, A., Levine, S.S., Cole, M.F., Frampton, G.M., Brambrink, T., Johnstone, S., Guenther, M.G., Johnston, W.K., Wernig, M., Newman, J., Calabrese, J.M., Dennis, L.M., Volkert, T.L., Gupta, S., Love, J., Hannett, N., Sharp, P.A., Bartel, D.P., Jaenisch, R., Young, R.A., 2008. Connecting microRNA genes to the core transcriptional regulatory circuitry of embryonic stem cells. Cell 134 (3), 521–533.

Martello, G., Zacchigna, L., Inui, M., Montagner, M., Adorno, M., Mamidi, A., Morsut, L., Soligo, S., Tran, U., Dupont, S., Cordenonsi, M., Wessely, O., Piccolo, S., 2007. MicroRNA control of nodal signalling. Nature 449 (7159), 183–188.

Massirer, K.B., Pasquinelli, A.E., 2006. The evolving role of microRNAs in animal gene expression. Bioessays 28 (5), 449–452.

Melkman-Zehavi, T., Oren, R., Kredo-Russo, S., Shapira, T., Mandelbaum, A.D., Rivkin, N., Nir, T., Lennox, K.A., Behlke, M.A., Dor, Y., Hornstein, E., 2011. miRNAs control insulin content in pancreatic beta-cells via downregulation of transcriptional repressors. EMBO J. 30 (5), 835–845.

Melton, C., Blelloch, R., 2010. MicroRNA regulation of embryonic stem cell self-renewal and differentiation. Adv. Exp. Med. Biol. 695, 105–117.

Melton, C., Judson, R.L., Blelloch, R., 2010. Opposing microRNA families regulate self-renewal in mouse embryonic stem cells. Nature 463 (7281), 621–626.

Morin, R.D., O'Connor, M.D., Griffith, M., Kuchenbauer, F., Delaney, A., Prabhu, A.L., Zhao, Y., McDonald, H., Zeng, T., Hirst, M., Eaves, C.J., Marra, M.A., 2008. Application of massively parallel sequencing to microRNA profiling and discovery in human embryonic stem cells. Genome Res. 18 (4), 610–621.

Murchison, E.P., Partridge, J.F., Tam, O.H., Cheloufi, S., Hannon, G.J., 2005. Characterization of Dicer-deficient murine embryonic stem cells. Proc. Natl. Acad. Sci. U. S. A. 102 (34), 12135–12140.

O'Loghlen, A., Munoz-Cabello, A.M., Gaspar-Maia, A., Wu, H.A., Banito, A., Kunowska, N., Racek, T., Pemberton, H.N., Beolchi, P., Lavial, F., Masui, O., Vermeulen, M., Carroll, T., Graumann, J., Heard, E., Dillon, N., Azuara, V., Snijders, A.P., Peters, G., Bernstein, E., Gil, J., 2012. MicroRNA regulation of Cbx7 mediates a switch of Polycomb orthologs during ESC differentiation. Cell Stem Cell 10 (1), 33–46.

Ouziel-Yahalom, L., Zalzman, M., Anker-Kitai, L., Knoller, S., Bar, Y., Glandt, M., Herold, K., Efrat, S., 2006. Expansion and redifferentiation of adult human pancreatic islet cells. Biochem. Biophys. Res. Commun. 341 (2), 291–298.

Packer, A.N., Xing, Y., Harper, S.Q., Jones, L., Davidson, B.L., 2008. The bifunctional microRNA miR-9/miR-9* regulates REST and CoREST and is downregulated in Huntington's disease. J. Neurosci. 28 (53), 14341–14346.

Pan, L., Gong, Z., Zhong, Z., Dong, Z., Liu, Q., Le, Y., Guo, J., 2011. Lin-28 reactivation is required for let-7 repression and proliferation in human small cell lung cancer cells. Mol. Cell. Biochem. 355 (1–2), 257–263.

Pasquinelli, A.E., Reinhart, B.J., Slack, F., Martindale, M.Q., Kuroda, M.I., Maller, B., Hayward, D.C., Ball, E.E., Degnan, B., Muller, P., Spring, J., Srinivasan, A., Fishman, M., Finnerty, J., Corbo, J., Levine, M., Leahy, P., Davidson, E., Ruvkun, G., 2000. Conservation of the sequence and temporal expression of let-7 heterochronic regulatory RNA. Nature 408 (6808), 86–89.

Paterson, E.L., Kolesnikoff, N., Gregory, P.A., Bert, A.G., Khew-Goodall, Y., Goodall, G.J., 2008. The microRNA-200 family regulates epithelial to mesenchymal transition. ScientificWorldJournal 8, 901–904.

Peyton, M., Stellrecht, C.M., Naya, F.J., Huang, H.P., Samora, P.J., Tsai, M.J., 1996. BETA3, a novel helix-loop-helix protein, can act as a negative regulator of BETA2 and MyoD-responsive genes. Mol. Cell Biol. 16 (2), 626–633.

Poy, M.N., Eliasson, L., Krutzfeldt, J., Kuwajima, S., Ma, X., Macdonald, P.E., Pfeffer, S., Tuschl, T., Rajewsky, N., Rorsman, P., Stoffel, M., 2004. A pancreatic islet-specific microRNA regulates insulin secretion. Nature 432 (7014), 226–230.

Poy, M.N., Hausser, J., Trajkovski, M., Braun, M., Collins, S., Rorsman, P., Zavolan, M., Stoffel, M., 2009. miR-375 maintains normal pancreatic alpha- and beta-cell mass. Proc. Natl. Acad. Sci. U. S. A. 106 (14), 5813–5818.

Qi, J., Yu, J.Y., Shcherbata, H.R., Mathieu, J., Wang, A.J., Seal, S., Zhou, W., Stadler, B.M., Bourgin, D., Wang, L., Nelson, A., Ware, C., Raymond, C., Lim, L.P., Magnus, J., Ivanovska, I., Diaz, R., Ball, A., Cleary, M.A., Ruohola-Baker, H., 2009. microRNAs regulate human embryonic stem cell division. Cell Cycle 8 (22), 3729–3741.

Raponi, M., Dossey, L., Jatkoe, T., Wu, X., Chen, G., Fan, H., Beer, D.G., 2009. MicroRNA classifiers for predicting prognosis of squamous cell lung cancer. Cancer Res. 69 (14), 5776–5783.

Redmer, T., Diecke, S., Grigoryan, T., Quiroga-Negreira, A., Birchmeier, W., Besser, D., 2011. E-cadherin is crucial for embryonic stem cell pluripotency and can replace OCT4 during somatic cell reprogramming. EMBO Rep. 12 (7), 720–726.

Reinhart, B.J., Slack, F.J., Basson, M., Pasquinelli, A.E., Bettinger, J.C., Rougvie, A.E., Horvitz, H.R., Ruvkun, G., 2000. The 21-nucleotide let-7 RNA regulates developmental timing in Caenorhabditis elegans. Nature 403 (6772), 901–906.

Rosa, A., Brivanlou, A.H., 2009. microRNAs in early vertebrate development. Cell Cycle 8 (21).

Rosa, A., Brivanlou, A.H., 2011. A regulatory circuitry comprised of miR-302 and the transcription factors OCT4 and NR2F2 regulates human embryonic stem cell differentiation. EMBO J. 30 (2), 237–248.

Rosa, A., Spagnoli, F.M., Brivanlou, A.H., 2009. The miR-430/427/302 family controls mesendodermal fate specification via species-specific target selection. Dev. Cell 16 (4), 517–527.

Rosati, J., Spallotta, F., Nanni, S., Grasselli, A., Antonini, A., Vincenti, S., Presutti, C., Colussi, C., D'Angelo, C., Biroccio, A., Farsetti, A., Capogrossi, M.C., Illi, B., Gaetano, C., 2011. Smad-interacting protein-1 and microRNA 200 family define a nitric oxide-dependent molecular circuitry involved in embryonic stem cell mesendoderm differentiation. Arterioscler. Thromb. Vasc. Biol. 31 (4), 898–907.

Rybak, A., Fuchs, H., Smirnova, L., Brandt, C., Pohl, E.E., Nitsch, R., Wulczyn, F.G., 2008. A feedback loop comprising lin-28 and let-7 controls pre-let-7 maturation during neural stem-cell commitment. Nat. Cell Biol. 10 (8), 987–993.

Sachdeva, M., Zhu, S., Wu, F., Wu, H., Walia, V., Kumar, S., Elble, R., Watabe, K., Mo, Y.Y., 2009. p53 represses c-Myc through induction of the tumor suppressor miR-145. Proc. Natl. Acad. Sci. U. S. A. 106 (9), 3207–3212.

Samavarchi-Tehrani, P., Golipour, A., David, L., Sung, H.K., Beyer, T.A., Datti, A., Woltjen, K., Nagy, A., Wrana, J.L., 2010. Functional genomics reveals a BMP-driven mesenchymal-to-epithelial transition in the initiation of somatic cell reprogramming. Cell Stem Cell 7 (1), 64–77.

Sampson, V.B., Rong, N.H., Han, J., Yang, Q., Aris, V., Soteropoulos, P., Petrelli, N.J., Dunn, S.P., Krueger, L.J., 2007. MicroRNA let-7a down-regulates MYC and reverts MYC-induced growth in Burkitt lymphoma cells. Cancer Res. 67 (20), 9762–9770.

Saunders, L.R., Sharma, A.D., Tawney, J., Nakagawa, M., Okita, K., Yamanaka, S., Willenbring, H., Verdin, E., 2010. miRNAs regulate SIRT1 expression during mouse embryonic stem cell differentiation and in adult mouse tissues. Aging (Albany NY) 2 (7), 415–431.

Schulman, B.R., Esquela-Kerscher, A., Slack, F.J., 2005. Reciprocal expression of lin-41 and the microRNAs let-7 and mir-125 during mouse embryogenesis. Dev. Dyn. 234 (4), 1046–1054.

Siemens, H., Jackstadt, R., Hunten, S., Kaller, M., Menssen, A., Gotz, U., Hermeking, H., 2011. miR-34 and SNAIL form a double-negative feedback loop to regulate epithelial-mesenchymal transitions. Cell Cycle 10 (24), 4256–4271.

Soldati, C., Bithell, A., Johnston, C., Wong, K.Y., Teng, S.W., Beglopoulos, V., Stanton, L.W., Buckley, N.J., 2012. Repressor element 1 silencing transcription factor couples loss of pluripotency with neural induction and neural differentiation. Stem Cells 30 (3), 425–434.

Subramanyam, D., Lamouille, S., Judson, R.L., Liu, J.Y., Bucay, N., Derynck, R., Blelloch, R., 2011. Multiple targets of miR-302 and miR-372 promote reprogramming of human fibroblasts to induced pluripotent stem cells. Nat. Biotechnol. 29 (5), 443–448.

Suh, M.R., Lee, Y., Kim, J.Y., Kim, S.K., Moon, S.H., Lee, J.Y., Cha, K.Y., Chung, H.M., Yoon, H.S., Moon, S.Y., Kim, V.N., Kim, K.S., 2004. Human embryonic stem cells express a unique set of microRNAs. Dev. Biol. 270 (2), 488–498.

Suh, N., Baehner, L., Moltzahn, F., Melton, C., Shenoy, A., Chen, J., Blelloch, R., 2010. MicroRNA function is globally suppressed in mouse oocytes and early embryos. Curr. Biol. 20 (3), 271–277.

Suzuki, H.I., Yamagata, K., Sugimoto, K., Iwamoto, T., Kato, S., Miyazono, K., 2009. Modulation of microRNA processing by p53. Nature 460 (7254), 529–533.

Tada, S., Era, T., Furusawa, C., Sakurai, H., Nishikawa, S., Kinoshita, M., Nakao, K., Chiba, T., 2005. Characterization of mesendoderm: a diverging point of the definitive endoderm and mesoderm in embryonic stem cell differentiation culture. Development 132 (19), 4363–4374.

Takahashi, K., Tanabe, K., Ohnuki, M., Narita, M., Ichisaka, T., Tomoda, K., Yamanaka, S., 2007. Induction of pluripotent stem cells from adult human fibroblasts by defined factors. Cell 131 (5), 861–872.

Takamizawa, J., Konishi, H., Yanagisawa, K., Tomida, S., Osada, H., Endoh, H., Harano, T., Yatabe, Y., Nagino, M., Nimura, Y., Mitsudomi, T., Takahashi, T., 2004. Reduced expression of the let-7 microRNAs in human lung cancers in association with shortened postoperative survival. Cancer Res. 64 (11), 3753–3756.

Takaya, T., Ono, K., Kawamura, T., Takanabe, R., Kaichi, S., Morimoto, T., Wada, H., Kita, T., Shimatsu, A., Hasegawa, K., 2009. MicroRNA-1 and microRNA-133 in spontaneous myocardial differentiation of mouse embryonic stem cells. Circ. J. 73 (8), 1492–1497.

Tang, X., Muniappan, L., Tang, G., Ozcan, S., 2009. Identification of glucose-regulated miRNAs from pancreatic {beta} cells reveals a role for miR-30d in insulin transcription. RNA 15 (2), 287–293.

Toker, A., Newton, A.C., 2000. Cellular signaling: pivoting around PDK-1. Cell 103 (2), 185–188.

Tsai, Z.Y., Singh, S., Yu, S.L., Kao, L.P., Chen, B.Z., Ho, B.C., Yang, P.C., Li, S.S., 2010. Identification of microRNAs regulated by activin A in human embryonic stem cells. J. Cell. Biochem. 109 (1), 93–102.

Tsukamoto, Y., Nakada, C., Noguchi, T., Tanigawa, M., Nguyen, L.T., Uchida, T., Hijiya, N., Matsuura, K., Fujioka, T., Seto, M., Moriyama, M., 2010. MicroRNA-375 is downregulated in gastric carcinomas and regulates cell survival by targeting PDK1 and 14-3-3zeta. Cancer Res. 70 (6), 2339–2349.

Tzur, G., Levy, A., Meiri, E., Barad, O., Spector, Y., Bentwich, Z., Mizrahi, L., Katzenellenbogen, M., Ben-Shushan, E., Reubinoff, B.E., Galun, E., 2008. MicroRNA expression patterns and function in endodermal differentiation of human embryonic stem cells. PLoS One 3 (11), e3726.

Ventura, A., Young, A.G., Winslow, M.M., Lintault, L., Meissner, A., Erkeland, S.J., Newman, J., Bronson, R.T., Crowley, D., Stone, J.R., Jaenisch, R., Sharp, P.A., Jacks, T., 2008. Targeted deletion reveals essential and overlapping functions of the miR-17 through 92 family of miRNA clusters. Cell 132 (5), 875–886.

Wang, J., Cao, N., Yuan, M., Cui, H., Tang, Y., Qin, L.J., Huang, X., Shen, N., Yang, H., 2012. MicroRNA-125b/Lin28 pathway contributes to the mesendodermal fate decision of embryonic stem cells. Stem Cells Dev.

Wang, Y., Baskerville, S., Shenoy, A., Babiarz, J.E., Baehner, L., Blelloch, R., 2008. Embryonic stem cell-specific microRNAs regulate the G1-S transition and promote rapid proliferation. Nat. Genet. 40 (12), 1478–1483.

Wang, Y., Medvid, R., Melton, C., Jaenisch, R., Blelloch, R., 2007. DGCR8 is essential for microRNA biogenesis and silencing of embryonic stem cell self-renewal. Nat. Genet. 39 (3), 380–385.

Wang, Z., Lin, S., Li, J.J., Xu, Z., Yao, H., Zhu, X., Xie, D., Shen, Z., Sze, J., Li, K., Lu, G., Chan, D.T., Poon, W.S., Kung, H.F., Lin, M.C., 2011. MYC protein inhibits transcription of the microRNA cluster MC-let-7a-1~let-7d via noncanonical E-box. J. Biol. Chem. 286 (46), 39703–39714.

Wienholds, E., Kloosterman, W.P., Miska, E., Alvarez-Saavedra, E., Berezikov, E., de Bruijn, E., Horvitz, H.R., Kauppinen, S., Plasterk, R.H., 2005. MicroRNA expression in zebrafish embryonic development. Science 309 (5732), 310–311.

Xu, N., Papagiannakopoulos, T., Pan, G., Thomson, J.A., Kosik, K.S., 2009. MicroRNA-145 regulates OCT4, SOX2, and KLF4 and represses pluripotency in human embryonic stem cells. Cell 137 (4), 647–658.

Yamakuchi, M., Ferlito, M., Lowenstein, C.J., 2008. miR-34a repression of SIRT1 regulates apoptosis. Proc. Natl. Acad. Sci. U. S. A. 105 (36), 13421–13426.

Yamakuchi, M., Lowenstein, C.J., 2009. MiR-34, SIRT1 and p53: the feedback loop. Cell Cycle 8 (5), 712–715.

Yu, J., Chau, K. F., Vodyanik, M. A., Jiang, J., and Jiang, Y., 2011. Efficient feeder-free episomal reprogramming with small molecules, PLoS One 6(3):e17557.

Yu, J., Li, A., Hong, S.M., Hruban, R.H., Goggins, M., 2012. MicroRNA alterations of pancreatic intraepithelial neoplasias. Clin. Cancer Res. 18 (4), 981–992.

Zhang, Z.W., Zhang, L.Q., Ding, L., Wang, F., Sun, Y.J., An, Y., Zhao, Y., Li, Y.H., Teng, C.B., 2011. MicroRNA-19b downregulates insulin 1 through targeting transcription factor NeuroD1. FEBS Lett. 585 (16), 2592–2598.

Zhao, B., Tumaneng, K., Guan, K.L., 2011. The Hippo pathway in organ size control, tissue regeneration and stem cell self-renewal. Nat. Cell Biol. 13 (8), 877–883.

Zhao, H., Zhu, L., Jin, Y., Ji, H., Yan, X., Zhu, X., 2012. miR-375 is highly expressed and possibly transactivated by achaete-scute complex homolog 1 in small-cell lung cancer cells. Acta Biochim. Biophys. Sin. (Shanghai) 44 (2), 177–182.

Zhong, X., Li, N., Liang, S., Huang, Q., Coukos, G., Zhang, L., 2010. Identification of microRNAs regulating reprogramming factor LIN28 in embryonic stem cells and cancer cells. J. Biol. Chem. 285 (53), 41961–41971.

Zhou, A.D., Diao, L.T., Xu, H., Xiao, Z.D., Li, J.H., Zhou, H., Qu, L.H., 2011. beta-Catenin/LEF1 transactivates the microRNA-371–373 cluster that modulates the Wnt/beta-catenin-signaling pathway. Oncogene.

Zovoilis, A., Smorag, L., Pantazi, A., Engel, W., 2009. Members of the miR-290 cluster modulate in vitro differentiation of mouse embryonic stem cells. Differentiation 78 (2–3), 69–78.

CHAPTER TWO

The Genetics of Dystonias

Mark S. LeDoux[*,†]
[*]Department of Neurology, University of Tennessee Health Science Center, Memphis, TN, USA
[†]Department of Anatomy and Neurobiology, University of Tennessee Health Science Center, Memphis, TN, USA

Contents

Advances in Genetics, Volume 79
ISSN 0065-2660,
http://dx.doi.org/10.1016/B978-0-12-394395-8.00002-5

Abstract

Dystonia has been defined as a syndrome of involuntary, sustained muscle contractions affecting one or more sites of the body, frequently causing twisting and repetitive movements or abnormal postures. Dystonia is also a clinical sign that can be the presenting or prominent manifestation of many neurodegenerative and neurometabolic disorders. Etiological categories include primary dystonia, secondary dystonia, heredodegenerative diseases with dystonia, and dystonia plus. Primary dystonia includes syndromes in which dystonia is the sole phenotypic manifestation with the exception that tremor can be present as well. Most primary dystonia begins in adults, and approximately 10% of probands report one or more affected family members. Many cases of childhood- and adolescent-onset dystonia are due to mutations in *TOR1A* and *THAP1*. Mutations in *THAP1* and *CIZ1* have been associated with sporadic and familial adult-onset dystonia. Although significant recent progress had been made in defining the genetic basis for most of the dystonia-plus and heredodegenerative diseases with dystonia, a major gap remains in understanding the genetic etiologies for most cases of adult-onset primary dystonia. Common themes in the cellular biology of dystonia include G1/S cell cycle control, monoaminergic neurotransmission, mitochondrial dysfunction, and the neuronal stress response.

I. INTRODUCTION

Although the term "dystonia" was coined in 1911 by Hermann Oppenheim, a leading German neurologist of his time, numerous earlier descriptions of dystonia had appeared in the medical literature under various names during the previous century (Fahn, 2011; Goetz *et al.*, 2001). Oppenheim believed that dystonia was an abnormality of tone with both hypertonic and hypotonic components. In 1836, J.H. Kopp eloquently described the clinical features of writer's cramp, a form of hand–forearm dystonia, in a German medical monograph. In his manual of diseases of the nervous system published in 1893, Sir William Richard Gowers used the term "tetanoid chorea" to describe what was apparently dystonia. In 1908, Schwalbe employed the term "tonic cramps" and described hereditary contributions to dystonia.

Under the heavy influence of Freud, dystonia was predominantly viewed as a psychiatric disorder for several decades in the early nineteenth century. This is not surprising given that truly psychogenic dystonia is relatively common and the often bizarre anatomical patterns, task specificity and gestes antagonistes characteristic of organic dystonia. In 1944, dystonia was resurrected as a true neurological disorder by Herz who used frame-by-frame cinematography to typify dystonic movements as slow, sustained, and forceful contortions of the trunk and limbs. Herz dropped "hypotonia" from the definition of dystonia, as initially applied by Oppenheim.

More recently (1984), dystonia was defined by an *ad hoc* committee of the Dystonia Medical Research Foundation (André Barbeau, Donald B. Calne, Stanley Fahn, C. David Marsden, John Menkes, and G. Frederick Wooten) as "a syndrome of sustained muscle contractions, frequently causing twisting and repetitive movements or abnormal postures." Although the clinical definition of dystonia has remained static over nearly 30 years, the classification of dystonia has evolved (Fahn, 2011). In 1976, Fahn and Eldridge provided an etiological classification of dystonia: primary (hereditary or sporadic), secondary (associated with heredodegenerative disease or environmental insults), or psychological. In the same year, David Marsden and colleagues (1976) published an anatomical classification of dystonia: focal, segmental, or generalized. In general, dystonia can be classified by etiology (primary or secondary), age of onset (<20 or >20 years), and anatomical distribution (focal, segmental, multifocal, hemidystonia, or generalized) (Fahn, 1987, 1988; Fahn *et al.*, 1998). More recent iterations and refinements of this basic classification scheme have been largely driven by developments in the genetics of dystonia and related neurological disorders (Fahn, 2011).

In some sense, genetics has biased the study of dystonia. Primary generalized dystonias usually affect children and a subset of cases is due to a ΔGAG mutation in Exon 5 of *TOR1A* (DYT1). *TOR1A* was the first gene to be causally associated with primary dystonia. In recent years, many patients with early-onset primary generalized DYT1 dystonia have responded dramatically to deep brain stimulation. Other "genetic" forms of dystonia such as DYT5, DYT11, and DYT12 also exhibit striking phenotypes and/or very positive responses to treatment. In reality, however, the vast majority of dystonia (>90%) seen by subspecialists in neurology clinics is late onset and focal or segmental in distribution (Xiao *et al.*, 2010). The genetics underpinnings of late-onset primary dystonia are only beginning to be unraveled (Xiao *et al.*, 2009, 2010, 2011, 2012).

II. CLINICAL FEATURES

A. Phenomenology and Demographics

Primary dystonia includes syndromes in which dystonia is the sole phenotypic manifestation with the exception that tremor can be present as well. Tremor is included in this definition since rhythmic activation of contiguous muscle groups can be seen in a significant fraction of patients with focal dystonias, particularly those with cervical or hand–forearm dystonia. Dystonia is characterized by (1) "abnormal" co-contraction of agonist and antagonist muscle groups, (2) "abnormal" prolongation of EMG bursts in muscles normally required for a specific motor act, (3) "impaired" volitional control of a group of somatotopically contiguous muscles, and (4) "impaired" inhibition of spinal and brainstem reflexes beyond the somatotopic extent of clinical involvement. "Abnormal" and "impaired" are not binary terms, which make it difficult to differentiate primary dystonia from psychogenic dystonia and other movement disorders based strictly on neurophysiological criteria. Most commonly, however, dystonia remains a purely clinical diagnosis made by experienced neurologists who know how to differentiate dystonia from other movement disorders. There are no definitive laboratory tests for primary dystonia—psychogenic dystonia has been reported in carriers of the DYT1 *TOR1A* ΔGAG mutation.

Overall, dystonia is more common in females (LeDoux, 2012a; Xiao *et al.*, 2010). For late-onset primary dystonia affecting the craniocervical musculature (e.g., blepharospasm, laryngeal dystonia, and cervical dystonia), the M:F ratio is 1:1.5–2. In contrast, the M:F ratio may be closer to 1:1 for hand–forearm dystonia including writer's cramp and other task-specific appendicular dystonias. For some genetically defined forms of dystonia such as DYT5 due to mutations in *GCH1*, penetrance is higher in females.

Dystonia, in general, and certain genetic etiologies are more common in certain ethnic or racial groups. For example, the *TOR1A* ΔGAG mutation is more common in Ashkenazi Jews than in other populations. In the United States, dystonia may be less common in African-Americans (Marras *et al.*, 2007), although this may reflect ascertainment bias (Puschmann *et al.*, 2011).

Up to 10% of patients with adult-onset primary dystonia may experience remissions, which are more common early in the disease course but may occasionally occur in long-standing dystonia (LeDoux, 2012a). Remissions may be permanent, but, more commonly, only last months or a few years, and have been reported in patients with a genetic diagnosis (e.g., *THAP1* dystonia). Sustained relief of motor symptoms has also been reported after cessation of GPi DBS. These findings suggest that dystonia may be the consequence of aberrant neural networks becoming trapped in local minima rather than irreversible cellular pathology (LeDoux *et al.*, 2012a).

Dystonia is typically mobile and abates with sleep. Dystonia is often precipitated by action or worsens with movement. A small percentage of patients with dystonia have "fixed dystonia," a term which has been the subject of controversy for many years. Fixed dystonia is often associated with peripheral trauma, complex regional pain syndrome, and psychiatric comorbidities. However, fixed dystonia can be the terminal consequence of long-standing, untreated primary "mobile" dystonia. For instance, in the years prior to botulinum toxin injections and deep brain stimulation, many patients with cervical dystonia progresses to fixed abnormal head postures.

Tremor is common in patients with dystonia and may be dystonic and/ or non-dystonic. Appendicular tremors, usually non-dystonic, are common in patients with craniocervical dystonia. Familial essential tremor is often associated with dystonia and may represent a distinct subtype of essential tremor (Hedera *et al.*, 2010).

Task specificity and sensory tricks (gestes antagonistes) are relatively unique features of the dystonias. Initially, task-specific hand–forearm dystonia may be extreme and only present when writing certain numbers or letters (Shamim *et al.*, 2011). Over time, task specificity may wane. Classic examples of task-specific dystonia include writer's cramp and embouchure dystonia. However, task-specific dystonia is not limited to the hand and face. For illustration, task-specific leg dystonia may only manifest when walking down steps, whereas walking on a level surface, up steps, and down steps backward are normal. Task-specific dystonias have been reported in golfers, typists, pianists, violinists, and croupiers. Sensory tricks are reported by more than 50% of patients with focal dystonia. Tricks are associated with transient improvement or resolution of dystonia. Classic examples include touching the chin, cheek, or occiput in cervical dystonia; and singing, touching the lateral brow, and looking downward in blepharospasm. In some patients, simply thinking about the trick (interoceptive stimulus) may help to alleviate their dystonia.

In general, the relationship between anatomical site of dystonia onset and age of onset follows a caudal-to-rostral gradient: blepharospasm (58 years), oromandibular dystonia (53 years), spasmodic dysphonia (46 years), cervical dystonia (45 years), hand–forearm dystonia (35 years), and distal leg dystonia (<20 years). Most commonly, distal leg dystonia begins in childhood with inversion and plantar flexion at the ankle and spreads rostrally. However, leg dystonia may occasionally appear in adults without foot inversion (Van Gerpen *et al.*, 2010). Site of onset may be gene specific. For example, DYT6/THAP1 dystonia usually begins in an arm or the neck, whereas DYT1 dystonia often begins in a leg.

Dystonia may spread from an initial site of onset (Weiss *et al.*, 2006). Risk for rostral spread is high in early-onset DYT1 dystonia that begins in a leg. In contrast, among the late-onset dystonias, risk of spread is highest for blepharospasm. Blepharospasm often spreads to the lower face and masticatory muscles and, in a smaller subset of patients, to the cervical musculature (LeDoux, 2009; Waln and LeDoux, 2011). Only rarely does blepharospasm spread to become generalized.

Patients with dystonia may experience pain and suffer from psychiatric comorbidities. Pain is perhaps most common in subjects with cervical dystonia but is also reported by individuals with blepharospasm, masticatory dystonia, and limb dystonia. Pain intensity may correlate poorly with the apparent severity of dystonic contractures. In some clinical series, subjects with primary focal dystonia are reported to have more obsessive-compulsive tendencies and a higher frequency of depressive disorders than matched control groups.

B. Classification

Dystonia can be classified by age of onset, distribution, and etiology (Table 2.1) (Fahn, 1987, 1988, 2011; Fahn *et al.*, 1998). In contrast to the presentation in Table 2.1 (Fahn, 1987, 1988; Fahn *et al.*, 1998), some of the more recently published classification schemes used 26 years of age as the dividing line between early- and late- or adult-onset dystonia (Fahn, 2011). However, based on normal patterns of human development, 20 years is a biologically rational separation point (Kuczmarski *et al.*, 2000). Application of 26 years arose from recommendations for DYT1 ΔGAG diagnostic testing (Bressman *et al.*, 2000) and is not applicable to non-DYT1 primary dystonia. In brief, DYT1 ΔGAG diagnostic testing was recommended for all individuals with primary dystonia with onset before age 26. Testing after age

Table 2.1 Classification of dystonia

Age of onset
Early onset (<20 years)
Late onset (>20 years)

Distribution

Focal: single body region (e.g., cervical dystonia, blepharospasm, oromandibular dystonia, spasmodic dysphonia, and task-specific dystonias)
Segmental: contiguous regions (e.g., cranial + cervical)
Multifocal: non-contiguous regions (e.g., cervical + leg)
Generalized: leg + trunk + one other body part
Hemidystonia: ipsilateral arm + leg

Etiology

Primary dystonia: syndromes in which dystonia is the sole phenotypic manifestation with the exception that tremor can be present as well
Secondary dystonia: due to structural lesions, neural insults, or medications (e.g., stroke, trauma, encephalitis, etc.)
Dystonia plus: dystonia plus another movement disorder without overt evidence of neurodegeneration (dopa-responsive dystonia [DRD/DYT5a and DYT5b], myoclonus-dystonia syndrome [MDS/DYT11], rapid-onset dystonia–parkinsonism [RDP/DYT12], and early-onset dystonia with parkinsonism [DYT16])
Heredodegenerative diseases with dystonia: dystonia may be a prominent feature (e.g., X-linked dystonia–parkinsonism [DYT3], progressive supranuclear palsy [PSP], Parkinson's disease, multiple system atrophy, corticobasal ganglionic degeneration, spinocerebellar ataxia type 3 [SCA3])
Paroxysmal dyskinesias: sudden episodes of involuntary movement, dystonia is often a major clinical feature
Psychogenic dystonia: dystonia is primarily due to psychological factors
Pseudodystonia: dystonia mimics associated with abnormal postures

26 may be considered for subjects having an affected relative with early onset. It should be emphasized that these guidelines only apply to DYT1 dystonia, which comprises less than 1% of all primary dystonia (Xiao *et al.*, 2009).

Age of onset is usually easy to establish and guides the clinician to underlying etiologies. For instance, DYT1 (Table 2.2) typically presents around 10 years of age with distal lower extremity dystonia. In contrast, the mean age of onset for primary focal dystonias of the head and neck is approximately 50 years (Xiao *et al.*, 2009, 2010). Among the more common focal dystonias are cervical dystonia (i.e., spasmodic torticollis), blepharospasm (abnormal contractions of the orbicularis oculi musculature), and task-specific hand–forearm dystonia (e.g., writer's cramp). The dystonia-plus

Table 2.2 Hereditary dystonias with Mendelian inheritance patterns

HUGO/OMIM	Common name	Locus/gene	Mode	Mutant protein
DYT1 (MIM 128100)	Oppenheim's dystonia	9q34.11/ *TOR1A*	AD	TorsinA (Ozelius *et al.*, 1997)
DYT2 (MIM 224500)	Autosomal recessive dystonia	Unknown	AR	Unknown
DYT3 (MIM 314250)	Lubag (X-linked dystonia–parkinsonism)	Xq13.1/*TAF1*	XLR	Reduced TAF1 expression (Makino *et al.*, 2007; Nolte *et al.*, 2003)
DYT4 (MIM 128101)	Australian whispering dysphonia family	Unknown	AD	Unknown
DYT5a (MIM 128230)	Dopa-responsive dystonia	14q22.2/ *GCH1*	AD	GTP cyclohydrolase I (Ichinose *et al.*, 1994)
DYT5b	Dopa-responsive dystonia	11p.15.5/*TH*	AR	Tyrosine hydroxylase (Lüdecke *et al.*, 1995)
DYT5b	Dopa-responsive dystonia	2q13.2/*SPR*	AR	Sepiapterin reductase (Bonafé *et al.*, 2001)
DYT6 (MIM 602629)	Mixed-type dystonia	8p11.21	AD	THAP1 (Fuchs *et al.*, 2009)
DYT7 (MIM 602124)	Familial torticollis	18p	AD	Unknown
DYT8 (MIM 118800)	Paroxysmal nonkinesigenic dyskinesia (PNKD)	2q35/*PNKD*	AD	Paroxysmal nonkinesigenic protein (Rainier *et al.*, 2004)
DYT9 (MIM 601042)	Paroxysmal choreoathetosis/spasticity	1p34.2/ *SLC2A1*	AD	Glucose transporter 1 (GLUT1) (Weber *et al.*, 2011)
DYT10 (MIM 128200)	Paroxysmal kinesigenic dyskinesia (PKD)	16p11.2/ *PRRT2*	AD	Proline-rich transmembrane protein 2 (Chen *et al.*, 2011)
DYT11 (MIM 159900)	Myoclonus-dystonia syndrome	7q21.3/*SGCE*	AD	ε-sarcoglycan (Zimprich *et al.*, 2001)

DYT12 (MIM 128235)	Rapid-onset dystonia—parkinsonism	19q13.2/ *ATP1A3*	AD	Na^+/K^+—ATPase α-3 subunit (de Carvalho Aguiar *et al.*, 2004)
DYT13 (MIM 607671)	Italian family-primary torsion dystonia	1p36.32-p36.13	AD	Unknown
DYT15 (MIM 607488)	Myoclonus dystonia, Canadian family	18p11	AD	Unknown
DYT16 (MIM 612067)	Young-onset dystonia—parkinsonism	2q31.2/ *PRKRA*	AR	Stress-response protein PRKRA (Camargos *et al.*, 2008)
DYT17 (MIM 612406)	Generalized dystonia with dysarthria and dysphonia	20p11.2-q13.12	AR	Unknown
DYT18 (MIM 61216)	Paroxysmal exertional dyskinesia associated with hemolytic anemia	1p34.2/ *SLC2A1*	AR	Glucose transporter 1 (GLUT1) (Weber *et al.*, 2008)
DYT19 (MIM 611031)	Paroxysmal kinesigenic dyskinesia (PKD)	16q13-q22.1	AD	Unknown
DYT20 (MIM 611147)	Paroxysmal nonkinesigenic dyskinesia 2 (PNKD2)	2q31	AD	Unknown
DYT21	Adult-onset mixed dystonia	2q14.3-q21.3	AD	Unknown

AD, autosomal dominant; AR, autosomal recessive.

category is superficially distinct from both the primary dystonias and heredodegenerative diseases with dystonia given that the definition of neurodegeneration is fuzzy and postmortem tissue for rigorous pathological analysis at the structural and ultrastructural levels has been limited. More problematic perhaps is the absence of parkinsonism in most cases of dopa-responsive dystonia (DRD) due to autosomal dominant mutations in GTP cyclohydrolase I, uncertain presence of parkinsonism in DYT16, and not infrequent absence of dystonia in individuals with mutations in *SGCE*.

GTP cyclohydrolase I is the rate-limiting enzyme in the synthesis of tetrahydrobiopterin. Tetrahydrobiopterin is a cofactor for tyrosine hydroxylase, tryptophan hydroxylase, and phenylalanine hydroxylase. Patients with DRD characteristically respond dramatically to very small doses of levodopa, a clinical finding that distinguishes DYT5 from the other hereditary dystonias. Secondary dystonias are due to acquired neural insults such as head trauma or stroke and can be distinguished from primary dystonias by the presence of additional abnormalities on neurological examination, magnetic resonance imaging (MRI) or computed tomographic imaging abnormalities, and their distinctive temporal profiles. Tardive dystonia is also classified as a secondary dystonia.

In many neurodegenerative diseases, dystonia may be either a prominent or presenting feature. Either cervical or appendicular dystonia may be present in up to one-third of patients with parkinsonian syndromes such as multiple system atrophy and progressive supranuclear palsy (PSP). However, in these patients, characteristic neurological and neuroimaging findings readily permit an accurate diagnosis. In particular, the presence of dementia, autonomic dysfunction, and/or oculomotor abnormalities in the hereditary and neurodegenerative diseases with dystonia sets these disorders apart from the primary dystonias. Family history, response to levodopa, and rate of disease progression provide additional, often useful, diagnostic clues.

C. Genetic Designations

Analysis of the individual DYT syndromes provides important insight into possible genetic and molecular mechanisms underlying the development of the late-onset primary focal/segmental dystonias, although the "DYT" nomenclature includes many syndromes that are not considered to be primary dystonias. In particular, DYT5, DYT11, and DYT14 are dystonia-plus syndromes, and patients with DYT8, DYT9, and DYT10 mutations frequently exhibit additional motor manifestations such as ataxia or chorea.

Review of Table 2.2 reveals that inheritance may occur in dominant, recessive (e.g., DYT2), or X-linked (DYT3) Mendelian patterns. The DYT2 and DYT17 loci may be responsible for sporadic cases of DYT1-negative childhood-onset dystonia. Autosomal recessive inheritance also suggests that haploinsufficiency may contribute to some adult-onset primary dystonias.

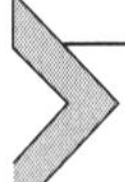

III. PRIMARY DYSTONIA

A. *TOR1A* (DYT1)

1. Genotypes and phenotypes

TOR1A (12 kb) is a five exon gene located on the reverse strand of Chr 9q34.11. The classic c.904_906delGAG (ΔGAG) mutation in Exon 5 of *TOR1A* typically manifests as early-onset generalized dystonia with onset in the distal lower extremities (Ozelius *et al.*, 1997). However, DYT1 dystonia is genetically and phenotypically heterogeneous. Although relatively uncommon, the DYT1 ΔGAG mutation has also been associated with late-onset focal, segmental, and multifocal dystonia (Gambarin *et al.*, 2006; Gasser *et al.*, 1998; Grundmann *et al.*, 2003; Kabakci *et al.*, 2004; O'Riordan *et al.*, 2002; Valente *et al.*, 1998). The carrier frequency of the classic DYT1 ΔGAG mutation is estimated at 1:1000–1:3000 in Ashkenazi Jews (Risch *et al.*, 1995) and less than 1:30,000 in non-Jews (Frédéric *et al.*, 2007). The penetrance of the DYT1 ΔGAG mutation is 30–40%. A p.Asp216His sequence variant may alter penetrance. Risch *et al.* (2007) found that the frequency of the 216His allele to be increased in ΔGAG mutation carriers without dystonia. Haplotype analysis demonstrated a protective effect of the His allele in trans with the ΔGAG mutation. Moreover, the Asp216 allele in cis may be required for penetrance of the ΔGAG mutation. The work of Risch *et al.* (2007) was not replicated in the French population (Frédéric *et al.*, 2009).

Another sequence variant in Exon 5 of *TOR1A* (c.863G>A, Fig. 2.1A) has been described in one female patient with severe childhood-onset generalized dystonia (Zirn *et al.*, 2008a). The G>A transition results in exchange of an arginine for a glutamine. Two additional sequence variants have been described in Exon 5. Leung *et al.* (2001) reported a subject with early-onset dystonia and myoclonus who harbored an 18-bp deletion in Exon 5 (p.Phe323_Tyr328del), which eliminates a putative phosphorylation site. The causality of the 18-bp deletion is doubtful since the same subject

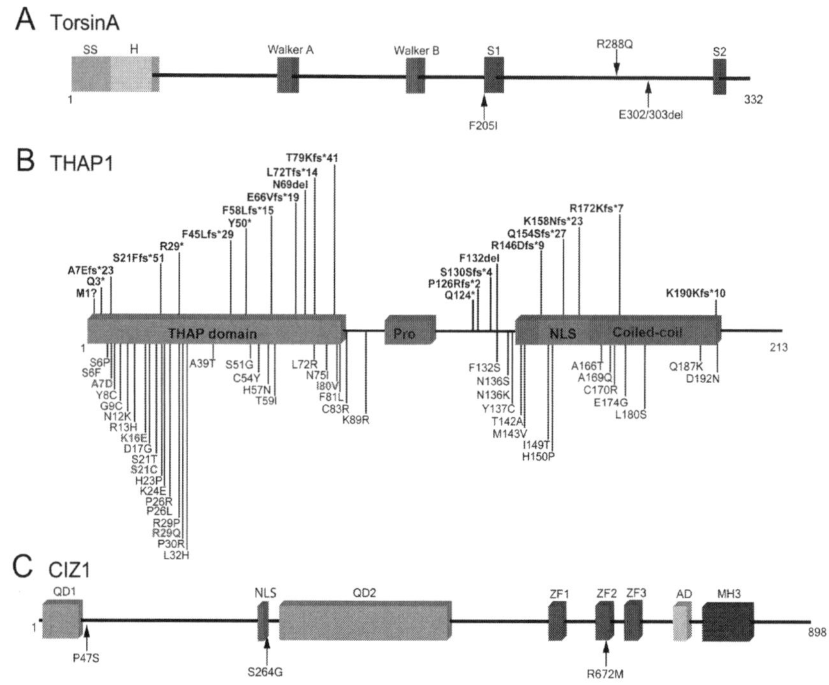

Figure 2.1 TorsinA, THAP1, and CIZ1. (A) Functional domains of torsinA and the location of coding variants that have been associated with dystonia. Walker A and B motifs are involved in ATP binding and hydrolysis. SS, signal sequence; H, hydrophobic domain; SI, sensor 1; and S2, sensor 2. (B) Functional domains of THAP1 and the location of coding variants that have been associated with dystonia. THAP, thanatos-associated protein domain; Pro, low-complexity proline-rich region; and NLS, nuclear localization signal. M1?, c.2delT or c.1A>G. (C) Functional domains of CIZ1 and the location of variants that have been associated with cervical dystonia. QD1, glutamine-rich domain 1; NLS, putative nuclear localization sequence; QD2, glutamine-rich domain 2; ZF, zinc finger domains; AD, acidic domain; and MH3, matrin (MATR3)-homologous domain 3. For color version of this figure, the reader is referred to the online version of this book.

was subsequently found to have a mutation in *SGCE*, the gene associated with myoclonus-dystonia syndrome (Klein *et al.*, 2002). In another study, a 4-bp deletion (c.934_937delAGAG) found in a putatively healthy blood donor should result in a premature stop at position 325 in the carboxy terminus of torsinA (Kabakci *et al.*, 2004).

A novel sequence variant (c.613T>A, p.F205I) in a patient with late-onset, masticatory and facial dystonia was reported by Calakos *et al.* (2010). The variant is located in a conserved AAA-ATPase domain of torsinA. Expression assays revealed that expression of p.F205I torsinA produced frequent intracellular inclusions. Despite functional data, the causality of this variant must be questioned given that the affected subject had a long history of psychiatric disease with neuroleptic exposure.

Xiao *et al.* (2009) sought to identify *TOR1A* Exon 5 mutations in a large cohort of subjects (>1000) with mainly non-generalized primary dystonia. High resolution melting (HRM) was used to examine the entire *TOR1A* Exon 5 coding sequence. HRM of Exon 5 showed high (100%) diagnostic sensitivity and specificity, reliably differentiating the ΔGAG and c.863G>A mutations. Melting curves were normal in 250/250 controls and 1012/1014 subjects with primary dystonia. The two subjects with shifted melting curves were found to harbor the classic ΔGAG deletion: (1) a non-Jewish Caucasian female with childhood-onset multifocal dystonia and (2) an Ashkenazi Jewish female with adolescent-onset spasmodic dysphonia. Although Exon 5 mutations in *TOR1A* are rarely associated with late-onset non-generalized primary dystonia, the role of sequence variants in Exons 1–4 and non-coding regions of *TOR1A* remains largely unknown. Clearly, the classic DYT1 ΔGAG mutation is uncommon in non-generalized primary dystonia and quite rare in late-onset primary dystonia.

Growing databases of sequence variants such as dbSNP, Exome Variant Server, and 1000 Genomes point out the richness of genetic variation and potential complexity of genotype–phenotype correlations. For instance, a substantial number of *TOR1A* variants have been reported in putatively normal controls (coding: p.Asp185His, p.Asp216His, p.Asp228His, and p.Asp259His; splice site; 5′UTR; 3′UTR; intronic; and promoter region). Unfortunately, however, most of the variants have not been validated with Sanger sequencing and many are probably next-gen read errors.

The characteristic DYT1 phenotype is characterized by early-onset in a limb, most commonly a leg, with spread to other limbs and the trunk over several years. Hand–forearm dystonia initially manifest as writer's cramp or more extensive upper limb dystonia without task specificity is the most common presentation of adolescent- or late-onset primary dystonia due to the DYT1 ΔGAG mutation, and most of these subjects will have a positive family history of dystonia (Gambarin *et al.*, 2006). However, DYT1 dystonia is phenotypically heterogeneous with oftentimes striking intrafamilial and interfamilial variability (Edwards *et al.*, 2003; Gajos *et al.*, 2007; Leube *et al.*, 1999; Opal *et al.*, 2002). Atypical presentations include (1) childhood-onset cervical dystonia with much later development of laryngeal dystonia and writer's cramp and (2) isolated laryngeal involvement for over 40 years (Xiao *et al.*, 2009). Other remarkable phenotypes described in the literature include onset of focal dystonia at 64 years, status dystonicus, and late-onset dystonia precipitated by exposure to a neuroleptic (Edwards *et al.*, 2003; Opal *et al.*, 2002).

2. TorsinA

TorsinA is a member of the ATPases associated with a variety of cellular activities (AAA+) superfamily of proteins that includes torsinB and two related gene products (TOR2A and TOR3A) in mammals, and OOC-5 in *Caenorhabditis elegans*. AAA+ proteins function as molecular chaperones for protein quality control (protein complex assembly, operation, disassembly, protein folding, unfolding, and degradation), membrane fusion and vesicular transport, and cytoskeletal regulation (Neuwald *et al.*, 1999; Ogura and Wilkinson, 2001; Vale, 2000). TorsinA harbors an N-terminal signal sequence, a single AAA+ module that includes Walker A and Walker B nucleotide-binding motifs, sensor 1 and sensor 2 regions, and two biochemically confirmed glycosylation sites (Callan *et al.*, 2007; Kamm *et al.*, 2004; Neuwald *et al.*, 1999; Ozelius *et al.*, 1997, 1999). TorsinA is an endoplasmic reticulum (ER) luminal monotopic membrane protein (Vander Heyden *et al.*, 2011). The hydrophobic N-terminal domain of torsinA directs static retention of torsinA within the ER by excluding it from ER exit sites. TorsinA functions as a homohexamer. TorsinA is enriched at the nuclear envelope when overexpressed *in vitro* (Goodchild and Dauer, 2004; Naismith *et al.*, 2004).

TorsinA is present in neuron perikarya and extends to the distal tips of dendrites and axons (Augood *et al.*, 2003; Kamm *et al.*, 2004; Konakova and Pulst, 2001; Konakova *et al.*, 2001). TorsinA may function as a chaperone for unfolded or degraded proteins and may facilitate movement of polytopic proteins to the cell surface (Torres *et al.*, 2004). TorsinA has been localized to Lewy bodies in Parkinson's disease brain and inclusion bodies in trinucleotide repeat diseases (Sharma *et al.*, 2001; Shashidharan *et al.*, 2000; Walker *et al.*, 2003). Overexpression of torsinA suppresses aggregation of α-synuclein in human neuroglioma cells (McLean *et al.*, 2002) and polyglutamine-induced protein aggregation in *C. elegans* (Caldwell *et al.*, 2003). TorsinA facilitates clearance of another dystonia-related protein, ε-sarcoglycan, by the ubiquitin proteosome system (Esapa *et al.*, 2007).

TorsinA appears to protect PC12 cells against cellular insults, such as serum deprivation and oxidative stress (Esapa *et al.*, 2007; Kuner *et al.*, 2003; Shashidharan *et al.*, 2004), and dopaminergic neurons from oxidative stress in mice (Kuner *et al.*, 2004) and *C. elegans* (Cao *et al.*, 2005). The chaperone functions of torsinA may be essential during developmental processes, which seemingly involve interaction with cytoskeletal elements (Ferrari-Toninelli *et al.*, 2004; Hewett *et al.*, 2006; Kamm *et al.*, 2004). The expression of torsinA shows prolonged increases after insults to the central nervous system (CNS) and peripheral nervous system (Zhao *et al.*, 2008). Moreover, expression of torsinA in reactive astrocytes in the CNS and satellite cells in the peripheral nervous system indicates that glial cells may contribute to the pathobiology of DYT1 dystonia (Zhao *et al.*, 2008).

The expression of torsinA is developmentally regulated with the highest levels of transcript and protein seen during the prenatal and early postnatal periods (Xiao *et al.*, 2004). The expression of torsinA is intense in cerebellar cortex and striatal cholinergic interneurons at Postnatal Day 14, a period of intense dendritogenesis in these areas (Vasudevan *et al.*, 2006; Xiao *et al.*, 2004).

The torsinA homolog present in *C. elegans* (OOC-5) contributes to PAR protein localization. Mutations of *ooc-5* result in polarity defects in *C. elegans* embryos (Basham and Rose, 2001). Attenuated torsinA expression promotes neurite outgrowth in SH-SY5Y human neuroblastoma cells (Ferrari-Toninelli *et al.*, 2004), whereas overexpression of mutant torsinA interferes with neurite extension (Hewett *et al.*, 2006). TorsinA may play direct or indirect roles in neuritogenesis and/or associated nuclear rotation.

Cytoskeletal interactions at the nuclear envelope and ER may be an important aspect of torsinA biology. TorsinA knockout and homozygous ΔGAG knockin mice show ultrastructural morphological abnormalities of the nuclear envelope (Goodchild *et al.*, 2005). TorsinA has been shown to interact with LAP1 in the nuclear envelope and LULL1 in the ER (Goodchild and Dauer, 2005). In a yeast two-hybrid study, torsinA was found to interact with kinesin light chain (Kamm *et al.*, 2004). TorsinA co-immunoprecipitates with a multimolecular complex that includes vimentin, tubulin, actin, kinesin light chain, LAP1, LULL1, and nesprin (Hewett *et al.*, 2006; Nery *et al.*, 2008) and related *in vitro* studies showed that mutant torsinA interferes with cytoskeletal events that involve vimentin (Hewett *et al.*, 2006). Vimentin, a member of the intermediate filament family of proteins, is expressed in developing brain (Hutchins and Casagrande, 1989; Sancho-Tello *et al.*, 1995) and reactive glia (Braun *et al.*, 1998; Kindy *et al.*, 1992).

B. DYT2

Initially described in Spanish gypsies (Gimenez-Roldan *et al.*, 1988; Santangelo, 1934), autosomal recessive generalized dystonia (DYT2; MIM 224500) has also been reported in Iranian (Khan *et al.*, 2003b) and Arab-American (Moretti *et al.*, 2005) families. In theory, DYT2 loci may be responsible for sporadic cases of DYT1- and DYT6-negative early-onset dystonia. Autosomal recessive inheritance also suggests that haploinsufficiency may contribute to some late-onset primary dystonias.

C. DYT4

The Australian "whispering dysphonia" kindred was first described by Parker (1985). Although the "whispering" descriptor suggests the presence of the abductor subtype, most affected individuals have reportedly presented with the adductor subtype of spasmodic dysphonia. DYT4 patients often progress to craniofacial and cervical dystonia. Some of the more severe cases progress to generalized dystonia. DYT4 appears to obey an autosomal dominant inheritance pattern. Studies of the Northern Queensland kindred have been confounded by the presence of Wilson's disease in several family members. However, *ATP7B* mutations do not segregate with dystonia (Wilcox *et al.*, 2011). Somewhat similar to DYT11, alcohol improves symptoms especially early in the disease course (Wilcox *et al.*, 2011). There is no linkage to the DYT1, DYT6, DYT7, DYT11, or DYT13 loci and no identified mutations in *THAP1* (DYT6) or *PRKRA* (DYT16).

D. *THAP1* (DYT6)

1. Genotype and phenotypes

THAP1 (THAP domain containing, apoptosis associated protein 1) was the second gene to be associated with primary dystonia (Fuchs *et al.*, 2009). The seminal *THAP1* mutation identified in Amish-Mennonite families was a heterozygous 5-bp (GGGTT) insertion followed by a 3-bp deletion (AAC) in Exon 2 (Fuchs *et al.*, 2009). At the protein level, this mutation causes a frameshift and premature stop codon (F45Lfs*29).

THAP1 is located on the reverse strand of Chr 8. *THAP1* consists of three exons and is alternatively spliced into three (2189 bp) and two (1993 bp) exon variants. *THAP1* mutations are an important cause of both early- and late-onset primary dystonia. In contrast to *TOR1A*, mutations in *THAP1* show great diversity with missense mutations broadly distributed across its three exons (Bressman *et al.*, 2009; LeDoux *et al.*, 2012b; Xiao *et al.*, 2010). In addition, frameshift, non-coding, and homozygous mutations in *THAP1* have also been associated with dystonia (Houlden *et al.*, 2010; Schneider *et al.*, 2011; Xiao *et al.*, 2010). Although initially described in Amish-Mennonites living in the Eastern United States, THAP1 dystonia has now been reported in individuals of Caucasians in most European countries and Chinese.

A promoter variant, 5′ to the *THAP1* coding sequence, was identified in African-Americans subjects with dystonia and African-American controls (Puschmann *et al.*, 2011). Highlighting the importance of matched control groups in candidate gene studies of dystonia, this variant (g.42698477C>T/A) was not found in Caucasians. Similarly, an initial report suggested that a nearby 5′UTR dinucleotide GA>TT variant was associated with increased risk for dystonia (Djarmati *et al.*, 2009), whereas a subsequent large, well-matched case–control study failed to confirm this association (Xiao *et al.*, 2011). However, the TT variant may be pathological in the homozygous state since all three published carriers had manifest dystonia (Djarmati *et al.*, 2009; Xiao *et al.*, 2011).

Initial reports suggested that c.71+126T>C (Houlden *et al.*, 2010) and c.71+9C>A (Xiao *et al.*, 2010) variants may increase risk for primary dystonia, and follow-up studies have been underpowered to substantiate or refute the pathogenicity of these variants (Djarmati *et al.*, 2009; Groen *et al.*, 2010; Lohmann *et al.*, 2012b). Although beyond the core splice site, the c.71+9C>A variant may possibly alter splicing efficiency and the ratio of the two *THAP1* isoforms. Similarly, the c.71+126T>C variant could exert cryptic effects on splicing or reduce overall gene expression.

At least 75 dystonia-associated coding variants have been published to date. Among this collection are 45 missense variants, 8 silent variants, and 22 indels or nonsense variants (Fig. 2.1B). All reported *THAP1* indels and nonsense mutations are predicted to elicit nonsense-mediated decay or generate truncated proteins. None of the mutations shown in Fig. 2.1B are predicted to generate an elongated THAP1. Missense variants are concentrated within the THAP domain. Other missense variants are present within the coiled-coil domain and nuclear localization signal. There are no missense variants within the proline-rich region. All but two missense variants (p.L32H and p.N136S) were heterozygous (Houlden *et al.*, 2010; Schneider *et al.*, 2011). The subject harboring p.N136S had no family history of dystonia, whereas three siblings exhibited early-onset generalized dystonia in the kindred identified with the p.L32H variant (Schneider *et al.*, 2011). Heterozygous carriers of the p.L32H variant were reportedly asymptomatic.

All reported silent variants are associated with sporadic dystonia. Silent mutations have the potential to be pathogenic if they activate cryptic splice sites, leading to exon skipping or the inclusion of intronic sequences into mature transcripts. For example, the c.267G>A (p.K89K) silent variant is located at the Exon 2 to Intron 2 boundary and was shown to reduce the expression of *THAP1* RNA in lymphocytes (Cheng *et al.*, 2011).

Age of onset for THAP1 dystonia ranges from 3 to over 60 years with a mean of 16.8 years (LeDoux *et al.*, 2012b). Common sites of onset are the arm (42.6%) and neck (24.6%), and 43.1% of affected individuals progress to generalized dystonia (LeDoux *et al.*, 2012b). Ultimately, the cranium, mainly the lower face, jaw, tongue, or pharynx, is affected in approximately half of patients with THAP1 dystonia. Mutations near the N-terminus of THAP1 are associated with an earlier age of onset and tended to be associated with more extensive anatomical involvement (LeDoux *et al.*, 2012b). Mutations within the THAP domain are associated with an earlier age of onset than non-THAP domain mutations (LeDoux *et al.*, 2012b). THAP domain mutations are also associated with more extensive anatomical distributions but have no significant effect on site of onset.

Each THAP1 sequence variant probably exhibits a unique penetrance value that is modified by environmental factors and overall genetic background. The penetrance of the Amish-Mennonite indel is 50–60% (Fuchs *et al.*, 2009), whereas the penetrance for many of the published missense variants is probably much lower. Proposing specific values for penetrance is compromised by intrafamilial variability in age of onset and assigning affection status to subtle phenotypes.

2. *THAP1*

THAP1 and related family members are defined by their zinc-binding THAP (Thanatos [Greek god of death]-associated protein) domain that is found in a number of proteins involved in various aspects of transcriptional regulation, apoptosis, and cell cycle control. The human genome contains 12 THAP family members. The THAP domain contains a C2-CH zinc finger that is similar to the DNA-binding domain of the Drosophila P-element transposase (Roussigne *et al.*, 2003). The THAP domain (1-81aa) of THAP1 recognizes an 11 nucleotide target sequence (AGTACG**GGCA**A) (Clouaire *et al.*, 2005). THAP1 also contains a low-complexity proline-rich region and bipartite nuclear localization signal. THAP1 probably functions as a homodimer within a larger multimeric DNA-binding complex. Amino acid residues 154 to 166 within the coiled-coil domain are critical for dimerization of THAP1 (Sengel *et al.*, 2011). *THAP1* is widely expressed in the brain and extra-neural tissues, including whole blood, liver, kidney, skeletal muscle, thyroid gland, and prostate.

E. DYT7

Leube *et al.* (1996) performed linkage analysis on a large German family (Family K) with seven definitely affected individuals (6/7 with cervical dystonia and 1/7 with spasmodic dysphonia) and six possibly affected individuals. Of the affected subjects, one also had blepharospasm and another had hand–forearm dystonia. The distribution of affected subjects was most compatible with autosomal dominant inheritance. A maximal LOD score of 3.17 was assigned to the region telomeric to marker D18S1153 on Chr 18p. In follow-up work, Leube *et al.* (1997) suggested that many sporadic cases of dystonia from Northwest Germany inherited the same mutation as Family K from a common ancestor and narrowed the causal mutation to a 6-cm region close to marker D18S1098.

F. DYT13

Dystonia has been linked to Chr 1p36.13-36.32 in a large Italian kindred with 11 definitely affected members (Bentivoglio *et al.*, 2004; Valente *et al.*, 2001). Linkage analysis using an autosomal dominant model generated a maximum LOD score of 3.44 between the disease and marker D1S2667. Age at onset ranged from 5 to 43 years, and similar to DYT6 dystonia, site of onset occurred either in the craniocervical region or arms. Dystonia generalized in only two cases and was relatively mild compared to DYT1 generalized dystonia. Penetrance was incomplete (~58%).

G. DYT17

Using homozygosity mapping with 382 microsatellite markers, DYT17 was mapped to a 20.5-Mb interval on Chr 20 in a large consanguineous Lebanese family with three affected individuals (Chouery *et al.*, 2008). Of 270 genes in this interval, causal variants were excluded in 27 candidate genes. Dystonia became manifest with cervical involvement between 14 and 19 years of age in the three affected individuals. Dystonia became generalized in one subject.

H. DYT21

Holmgren *et al.* (1995) reported a large Swedish family with 10 affected members with age of onset ranging from 18 to 50 years. Dystonia was quite variable with focal and generalized distributions. Initial presentations included blepharospasm, arm dystonia, and dysarthria.

In this Swedish kindred, dystonia was inherited in an autosomal dominant manner with a penetrance of approximately 90%. Using SNP-based linkage analysis, the dystonia locus was mapped to Chr 2q14.3-q21.3. Microsatellite confirmation generated a maximum LOD score of 5.59 for marker D2S1260. No causal mutations were identified in 22 candidate genes within the disease interval on Chr 2p (Norgren *et al.*, 2011).

I. CIZ1

In very recent work (Xiao *et al.*, 2012), solution-based whole-exome capture and massively parallel sequencing in combination with microsatellite-based linkage analysis was used to identify the causal sequence variant in a large Caucasian pedigree with adult-onset primary cervical dystonia first described by Uitti and Maraganore (1993). Cervical dystonia showed strongest linkage to microsatellite marker D9S159 located on Chr 9q34.11, and whole-exome sequencing identified an exonic splicing enhancer mutation in Exon 7 of *CIZ1* (c.790A>G, p.S264G), a gene within the candidate region. *CIZ1* encodes a $p21^{Cip1/Waf1}$-interacting zinc finger protein (CIZ1) that is involved in DNA synthesis and G1/S cell cycle control. After confirmation of co-segregation with dystonia, high-throughput screening identified two additional *CIZ1* missense mutations (p.P47S and p.R672M) among a population of 308 Caucasians with familial or sporadic adult-onset cervical dystonia (Fig. 2.1C).

J. Other Adult-Onset Primary Dystonia

Although at least 10-fold more common than early-onset (<20 years) primary dystonia and far more prevalent than well-known neurological disorders such as Huntington disease, amyotrophic lateral sclerosis, and Duchenne muscular dystrophy, very little is known about the biological underpinnings of the primary adult-onset focal dystonias. Moreover, we do not know if focal dystonias such as cervical dystonia, spasmodic dysphonia, blepharospasm, and hand–forearm dystonia (e.g., writer's cramp) are (1) distinct and separate entities or (2) localized manifestations of primary generalized dystonia. Clinical information suggests that age, gender, and environment are major contributors to the anatomical distribution of dystonia. Alternatively, the primary adult-onset focal dystonias are caused by shared genetic variant(s) transmitted in autosomal dominant fashion with variable expressivity. Phenotypic concordance–discordance is approximately 50%/50% for adult-onset primary dystonia. An example of phenotypic discordance would be blepharospasm in a proband and cervical dystonia in one of the proband's siblings.

Current thinking suggests that sporadic primary dystonia, like many other adult-onset diseases, is due to complex interactions between the genome and environment. In this regard, peripheral trauma (e.g., ocular surface irritation) or intense sensorimotor training (e.g., writing) may trigger dystonia in genetically predisposed individuals. Clearly, genetic factors play a major role in adult-onset primary dystonia since 10–20% of patients with primary adult-onset dystonia have one or more family members affected with dystonia (Defazio *et al.*, 2011; Xiao *et al.*, 2010) and several of the early-onset primary dystonias inherited in Mendelian fashion begin focally, show incomplete penetrance, and exhibit variable anatomical expressivity. It is likely that rare sequence variants of low to moderate penetrance in *THAP1*, *TOR1A*, *CIZ1*, *PRKRA*, and within the DYT2, DYT4, DYT7, DYT13, DYT16, DYT17, and DYT21 loci contribute to risk for largely sporadic adult-onset primary dystonia.

IV. DYSTONIA PLUS

A. DYT5/DRD

First described by Segawa *et al.* (1976), DRD is something of a misnomer since several other neurogenetic disorders may exhibit dystonia that responds to levodopa. On the other hand, classic DRD due to mutations in *GCH1* is imminently treatable with low dosages of levodopa and, diagnostically, should not be missed by clinicians.

Classic DRD (DYT5a) is due to deficiency of GTP cyclohydrolase I (GCH1), the rate-limiting enzyme for synthesis of the cofactor tetrahydrobiopterin (BH4) (Werner *et al.*, 2011). BH4 is an essential co-factor for the aromatic amino acid hydroxylases tyrosine hydroxylase (*TH*), tryptophan hydroxylase (*TPH2*), and phenylalanine hydroxylase (*PAH*) (Thony and Blau, 2006). BH4 also regulates nitric oxide synthase. TH is the rate-limiting enzyme for synthesis of dopamine, norepinephrine, and epinephrine. Tryptophan hydroxylase is the rate-limiting enzyme involved in synthesis of serotonin. The synthesis of BH4 requires the serial action of GTP cyclohydrolase I (*GCH1*), 6-pyruvoyl-tetrahydropterin synthase (*PTS*), and sepiapterin reductase (*SPR*). Autosomal recessive mutations in *PTS* are associated with hyperphenylalaninemia and can cause a wide range of neurological manifestations including microcephaly, oculogyric crises, sialorrhea, developmental delay, parkinsonism, seizures, and dystonia (Dudesek *et al.*, 2001).

Cofactor regeneration requires pterin-4a-carbinolamine dehydratase (*PCBD1*) and quinoid dihydropteridine reductase (*QDPR*) (Werner *et al.*, 2011). Recessive mutations in QDPR are associated with hyperphenylalaninemia, development delay, intracerebral calcifications, seizures, and, occasionally, dystonia (Larnaout *et al.*, 1998). In contrast, recessive mutations in *PCBD1* cause hyperphenylalaninemia but only mild, transient neurological features.

1. DYT5a (GTP cyclohydrolase I)

GCH1 contains six exons and is located on Chr 14q22.2. Grotzsch *et al.* (2002) mapped another locus for DRD (DYT14) to Chr 14q13. However, Wider *et al.* (2008) showed that this Swiss family with DRD had a heterozygous deletion of *GCH1* Exons 3 to 6. The vast majority of dominant mutations in *GCH1* are associated with the classic phenotype of childhood- and limb-onset dystonia that is completely responsive to low-dose levodopa (<400 mg/day). Transient levodopa-induced dyskinesias may be seen in some patients. Mean age of onset is approximately 6 years. The penetrance is notably higher in females.

Over time, the phenotype of GTP cyclohydrolase I–deficient DRD has expanded to include adult-onset parkinsonism, adult-onset focal dystonias including cervical dystonia, spastic diplegia, or cerebral palsy like syndrome, and a variety of co-morbidities including depression, anxiety, and sleep disorders. Upon careful examination, many untreated patients with childhood-onset DRD show evidence of mild parkinsonism with rigidity, bradykinesia, and rapid fatiguing with repetitive arm movements. Most patients report diurnal fluctuation of symptoms with worsening in the late afternoon hours.

The majority of GCH1 mutations can be detected with Sanger sequencing. However, a significant minority of affected families harbor large deletions or duplications or mutations in noncoding regions (Sharma *et al.*, 2011; Zirn *et al.*, 2008b). Therefore, most commercial laborites employ the quantitative polymerase chain reaction or multiplex ligation-dependent probe amplification if routine sequencing is unrevealing.

Rarely, GTP cyclohydrolase I deficiency is recessive. These cases are commonly associated with hyperphenylalaninemia and severe neurological dysfunction including seizures, developmental delay, and involuntary movements. These severe cases demand treatment with BH4, levodopa, and

5-hydroxytryptophan. Milder variants without hyperphenylalaninemia have also been reported and may respond to levodopa alone (Horvath *et al.*, 2008; Opladen *et al.*, 2011).

2. DYT5b (tyrosine hydroxylase)

TH deficiency shows a much broader phenotypic spectrum than GTP cyclohydrolase I deficiency (Willemsen *et al.*, 2010). At one end of the spectrum is mild dystonia that responds to low-dose levodopa for decades. In contrast, more severe cases are associated with infantile parkinsonism with onset prior to 6 months of age, developmental delay, levodopa-induced dyskinesias, ptosis, gastroparesis, and oculogyric crises. Although the vast majority of cases are autosomal recessive, heterozygotes may manifest subtle exertion-induced dystonia or rigidity or restless legs syndrome (Swoboda *et al.*, 2006). Most reported variants have been missense mutations, either homozygous or heterozygous. Frameshift and promoter region mutations have also been described (Verbeek *et al.*, 2007). The promoter mutations have been localized to a cAMP response element.

3. DYT5b (sepiapterin reductase)

In general, the lack of hyperphenylalaninemia distinguishes sepiapterin reductase deficiency from other autosomal recessive disorders of BH4 synthesis (Bonafé *et al.*, 2001). On average, clinical manifestations are more severe than those associated with GTP cyclohydrolase I deficiency. The disease spectrum includes cognitive delay, oculogyric crises, microcephaly, hyperactivity, hypersomnolence, parkinsonism, dystonia, tremor, seizures, and, if untreated, mental retardation. Onset may occur in infancy. Several novel homozygous and compound heterozygous missense, splice site, nonsense, and frameshift mutations in *SPR* have been reported since 2001 (Lohmann *et al.*, 2012a). In rare cases, sepiapterin reductase deficiency may be autosomal dominant (Steinberger *et al.*, 2004).

B. DYT11/MDS

Myoclonus-dystonia syndrome is due to mutations in *SGCE*, which encodes ε-sarcoglycan (Zimprich *et al.*, 2001). *SGCE* is maternally imprinted and covers 71 kb on the reverse strand of Chr 7q21.3. Virtually, all affected individuals have myoclonus, which is concentrated in the upper extremities, neck, and trunk with infrequent involvement of the legs. Approximately 50–65% of patients have dystonia, usually affecting the neck or arms. Very rarely, dystonia may be the only disease manifestation. Onset is usually in childhood or early adolescence but has been reported in the fourth decade of life. Many affected individuals reported a dramatic reduction in myoclonus after consumption of alcohol. Psychiatric disorders including depression and anxiety can be prominent non-motor features of the disorder.

Given that *SGCE* is maternally imprinted, penetrance is significantly higher when the mutant allele is inherited from the father. DYT11 may be caused by a variety of mutations in *SGCE* (nonsense, missense, insertions, and deletions). Testing for exonic deletions should be considered in individuals with a classic phenotype in whom Sanger sequencing is unrevealing (Asmus *et al.*, 2005; Grunewald *et al.*, 2008).

In muscle, the sarcoglycans are part of the large transmembrane dystrophin–glycoprotein complex required for the stability of striated muscle membranes. Muscle expression of ε-sarcoglycan is highest in the early postnatal period with only minimal expression in adult muscle (Xiao and LeDoux, 2003). In brain, ε-sarcoglycan is expressed at high levels in cerebellar cortex (Xiao and LeDoux, 2003), and a major brain-specific isoform shows relatively high expression in Purkinje cells and the dentate nucleus (Ritz *et al.*, 2011). In neurons, the dystrophin–glycoprotein complex may be involved in the clustering and stabilization of GABAergic synapses (Waite *et al.*, 2009).

C. DYT12/RDP

Although denoted "rapid-onset dystonia–parkinsonism (RDP)," relatively few patients have all four cardinal features of Parkinson's disease (resting tremor, bradykinesia, postural instability, and rigidity). In fact, the term "rapid-onset dystonia" would perhaps be more appropriate if it was not for the fact that many patients with largely sporadic adult-onset dystonia wake up with their disorder or report onset over a couple of days. Despite these caveats, RDP is fairly distinct with abrupt onset of dystonia with varying degrees of bradykinesia. The dystonia obeys a rostral-caudal gradient (face → arm → trunk) with minimal leg involvement and often prominent bulbar involvement with dysphagia and dysarthria. Patients do not respond to levodopa. Typically, RDP first manifests in teenagers or young adults, although age of onset ranges from 4 to 55 years of age.

RDP is an autosomal-dominant disorder due to mutations in *ATP1A3*, which encodes ATPase, Na^+/K^+ transporting, and α-3 polypeptide. The α-3 subunit of Na^+/K^+–ATPase is an integral membrane protein and a member of the P-type cation transport ATPases. Members of this family help to maintain electrochemical gradients across the plasma membrane. The α subunit of Na^+/K^+–ATPase has three isoforms (α-1, α-2, and α-3) encoded by distinct genes rather than alternative splicing of a single gene.

ATP1A3 contains 23 exons and covers 27.6 kb of genomic DNA. Mutations described to date have been concentrated in Exons 8, 14, 15, 17, 20, and 23 (Brashear *et al.*, 2007). Missense mutations, a 3-bp deletion and a 3-bp insertion, have been reported, mainly in Caucasians (Brashear *et al.*, 2007; de Carvalho Aguiar *et al.*, 2004; Tarsy *et al.*, 2010). An RDP p.D923N mutation in *ATP1A3* shows approximately 200-fold reduction of Na^+ affinity for activation of phosphorylation from ATP (Einholm *et al.*, 2010). In general, RDP mutant forms of Na^+/K^+ ATPase show reduced affinity for cytoplasmic Na^+ (Rodacker *et al.*, 2006).

RDP is genetically heterogeneous. *ATP1A3* mutations have not been identified in all subjects with an RDP phenotype (Brashear *et al.*, 2007). In particular, Kabakci *et al.* (2005) reported a large RDP kindred that did not harbor an *ATP1A3* mutation or show linkage to Chr 19.

D. DYT16

High-density autozygosity mapping (Illumina HumanHap550) was used to identify a disease-segregating region of homozygosity in affected family members from two families with early onset, apparently autosomal recessive dystonia in southeast-central Brazil (Camargos *et al.*, 2008). The relationship between these two families remains unclear. Age of onset ranged from 2 to 18 years. Lower limb onset was apparent in 4/7 subjects. Dystonia became generalized with neck, trunk, laryngeal, and oromandibular involvement. Some patients were mildly bradykinetic and possibly parkinsonian. Moreover, 3/7 subjects had evidence of upper motor neuron dysfunction. Sanger sequencing of all genes within a 1.2-Mb interval on Chr 2q31 revealed a single disease-segregating mutation, c.665C>T (P222L), in *PRKRA*, which encodes protein kinase, interferon-inducible double-stranded RNA-dependent activator. In a single follow-up study, Seibler *et al.* (2008) reported a heterozygous frameshift mutation in a patient with early-onset dystonia first manifest in a leg. At last examination, this subject had generalized dystonia that spared the cranial and facial muscles. The pathogenicity of this variant must be questioned given that *PRKRA*-null mice do not show evidence of dystonia or significant neurological disease (Peters *et al.*, 2009).

Furthermore, in contrast to the Brazilian patients, dystonia spared the cranial muscles in the subject from Germany.

V. HEREDODEGENERATIVE DYSTONIA

A. X-Linked Dystonia–Parkinsonism (XDP/DYT3)

Also known as Lubag, X-linked dystonia–parkinsonism (XDP) is limited to Filipinos with origin from the Island of Panay. Parkinsonism is usually the presenting manifestation in the fourth decade of life. Craniocervical dystonia typically begins later with involvement of the masticatory, cervical, and upper facial muscles in many patients. Appendicular dystonia is less common. Parkinsonism may partially respond to levodopa and dopamine agonists. Males are affected more severely than females and moderate phenotypic variability is noted with isolated parkinsonism or focal dystonia in some subjects.

Subjects with XDP harbor five distinct single nucleotide sequence variants in addition to a 48-bp deletion and an SVA (short interspersed nuclear element, variable number of tandem repeats, and *Alu* composite; SINE/VNTR/*Alu*) element (Makino *et al.*, 2007; Nolte *et al.*, 2003). The SVA is located in Intron 32 of TATA-binding protein-associated factor 1 gene (*TAF1*) and may alter expression of a neuron-specific isoform of TAF1 (N-TAF1). In rat brain, N-TAF1 is expressed in medium spiny neurons, preferentially in the striosome compartment of the striatum (Sako *et al.*, 2011). This finding is consistent with postmortem neuropathology in human subjects, which is characterized by a mosaic pattern of striatal gliosis with more prominent neuronal loss in striosomes than in the matrix (Evidente *et al.*, 2002; Goto *et al.*, 2005).

B. Parkinson's Disease

Dystonia is very common in sporadic and familial Parkinson's disease (Tolosa and Compta, 2006). Dystonia is a frequent "side-effect" of pharmacological and surgical treatment of Parkinson's disease. Well-recognized manifestations include AM off-dystonia in the lower extremities, blepharospasm with apraxia of eyelid opening, cervical dystonia, peak-dose dystonia, and diphasic dystonia.

Dystonia is often a presenting sign in early-onset Parkinson's disease. Classically, this manifests as action dystonia in a distal leg, occasionally only precipitated by exercise (Khan *et al.*, 2003a). Over 30% of individuals with recessive mutations in *PARK2*, which encodes parkin, present with leg dystonia. Distal leg dystonia is also a common presentation of recessive mutations in PARK6 (*PINK1*) and PARK7 (*DJ1*) mutations (Schneider and Bhatia, 2010a).

C. Tauopathies

Largely based on postmortem pathological identification of tau protein aggregation, a group of neurodegenerative disorders including PSP, corticobasal ganglionic degeneration (CBGD), frontotemporal dementia and parkinsonism linked to Chr 17 (FTDP-17), argyrophilic grain disease, and Pick's disease have been commonly referred to as tauopathies. Dystonia is a common clinical manifestation in PSP (Barclay and Lang, 1997) and CBGD (Reich and Grill, 2009). Tau is encoded by *MAPT*, which covers 134 kb on Chr 17q21.31. Tau is mainly expressed in neurons and contributes to the assembly and stabilization of microtubules. Alternative splicing of Exon 10 determines the number of microtubule-binding repeats (3R or 4R). In PSP and CBGD, accumulation of 4R-tau predominates, whereas 3R-tau is deposited in Pick's disease.

Case–control studies have demonstrated significant associations between the H1 haplotype, an inversion polymorphism of *MAPT* and risk for PSP and CGBD (Baker *et al.*, 1999; Houlden *et al.*, 2001). Missense mutations in *MAPT* have also been associated with sporadic and familial PSP (e.g., p.R5L and p.G303V) (Choumert *et al.*, 2011; Poorkaj *et al.*, 2002). There are no reported relationships among *MAPT* haplotypes and sporadic primary dystonia. Rare cases of CGBD have been associated with mutations in *PGRN*, which encodes progranulin (Spina *et al.*, 2007). *PGRN* mutations are more commonly associated with a clinical–pathological phenotype sometimes described as behavior-predominant frontotemporal dementia.

D. Neurodegeneration with Brain Iron Accumulation

Dystonia may be a presenting or prominent clinical feature of the neurodegeneration with brain iron accumulations (NBIAs), which are characterized by iron deposition in the brain, particularly the basal ganglia (Schneider and Bhatia, 2010b). With the advent of MRI and increased use of T2* and T2 fast-spin echo sequences, these disorders are being recognized with increasing frequency (McNeill *et al.*, 2008). It should be emphasized, however, that iron deposition in the basal ganglia is a part of normal aging, and increased iron deposition has been associated with numerous acquired and hereditary disorders that affect the CNS (Aquino *et al.*, 2009).

The classic disorder in this family is NBIA1 (*PANK2*, pantothenate kinase 2), previously called Hallervorden–Spatz disease. Other NBIAs and NBIA-like disorders include NBIA2 (*PLA2G6*; calcium-dependent cytosolic phospholipase A2, group VI) also known as PARK4 or infantile neuroaxonal dystrophy (INAD); NBIA3 (*FTL*, ferritin light chain) or neuroferritinopathy; NBIA4 (*C19orf12*, an orphan mitochondrial protein); aceruloplasminemia (*CP*, ceruloplasmin); FAHN (fatty acid hydroxylase neurodegeneration) disease or SPG35 (*FA2H*, fatty acid 2-hydroxylase); and Kufor–Rakeb syndrome (*ATP13A2*, lysosomal type 5 ATPase), also known as PARK9. With the exception of neuroferritinopathy, all of these disorders are autosomal recessive.

VI. DYSTONIA IN ASSOCIATION WITH OTHER NEUROGENETIC DISORDERS

A. Paroxysmal Dyskinesias

This group of neurological disorders is characterized by the sudden onset of involuntary movements that may include one or more of the following: dystonia, chorea, athetosis, and ballism (Bhatia, 2011). The paroxysmal dyskinesias are divided into four major types: paroxysmal kinesigenic dyskinesia (PKD), paroxysmal nonkinesigenic dyskinesia (PNKD), paroxysmal exertion-induced dyskinesia (PED), and paroxysmal hypnogenic dyskinesia. Precipitating factors are included within the name of each type: kinesigenic (sudden movements), nonkinesigenic (may occur spontaneously but often triggered by stressors such as anxiety, temperature extremes, or caffeine), exertion induced (running or prolonged walking), and hypnogenic (non-rapid eye movement sleep). The paroxysmal dyskinesias may be familial (hereditary), sporadic, or secondary (or symptomatic) to other neurological or metabolic disorders (e.g., head trauma, multiple sclerosis, hypoparathyroidism). Overall, genetic etiologies have been identified for a modest number affected families and small number of sporadic/idiopathic cases.

1. DYT8 (PNKD)

Many cases of primary, familial PNKD are due to missense mutations (p.A7V, p.A9V, and p.A33P) in the N-terminal mitochondrial targeting sequence of myofibrillogenesis regulator-1 or PNKD (MR-1 or PNKD;

Bruno *et al.*, 2007; Ghezzi *et al.*, 2009; Lee *et al.*, 2004; Rainier *et al.*, 2004). Patients with PNKD due to mutations in *PNKD* have attacks that last from 10 min to 1 h, often precipitated by caffeine, alcohol, or emotional stress. Onset occurs in infancy or early childhood and typically includes a combination of chorea and dystonia with involvement of the face, trunk, and limbs, often bilaterally.

PNKD shows homology to glyoxalase II and is probably involved in the cellular stress response. Although the primary substrate of PNKD is not known, glyoxalase II catalyzes the second step in the glutathione-dependent glyoxylase pathway, which converts methylglyoxal to D-lactic acid. Lower glutathione levels have been noted in the brains of PNKD knockout mice (Shen *et al.*, 2011).

2. DYT9/DYT18 (PED)

Recent work has shown that DYT9 (paroxysmal choreoathetosis/spasticity) and DYT18 (paroxysmal exercise-induced dyskinesia) are allelic disorders due to mutations in *SLC2A1*, which encodes the glucose transporter type 1 (GLUT1) (Weber and Lerche, 2009; Weber *et al.*, 2011).

Auburger *et al.* (1996) described a large pedigree with autosomal dominant "paroxysmal choreoathetosis/spasticity." The pedigree included 18 affected family members. Age of onset ranged from 2 to 15 years with episodes of involuntary movements lasting less than 20 min and manifest mainly as dystonia. The frequency of episodes was highly variable and ranged from daily to twice yearly. Episodes were triggered by a broad range of triggers including emotional stress, exercise, lack of sleep, and alcohol. Some family members demonstrated interictal spasticity in the lower extremities. The causal gene was mapped to Chr 1p.

Weber and co-workers (Schneider *et al.*, 2009; Weber *et al.*, 2008) showed that paroxysmal-exertion (or exercise)-induced dyskinesias (PED) were due to mutations in *SLC2A1*. They described a large three-generation family with PED, epilepsy, developmental delay, and hemolytic anemia. Cerebrospinal glucose levels were also reduced. Mutations in *SLC2A1* have also been identified in sporadic PED (Schneider *et al.*, 2009). In subsequent work, the kindred initially described by Auburger *et al.* (1996) and an Australian monozygotic twin pair were found to have causal mutations in *SLC2A1* with decreased glucose uptake in functional assays.

The classical GLUT1 deficiency phenotype is characterized by early-onset cognitive and motor delay, intractable epilepsy, microcephaly, involuntary movements (dystonia, chorea, ataxia, and choreoathetosis), spasticity, and microcephaly (Verrotti *et al.*, 2012). Most patients with GLUT1 syndromes have abnormally low levels of cerebrospinal fluid glucose. The most effective treatment for these disorders is a ketogenic diet (Verrotti *et al.*, 2012).

GLUT1 is expressed in the brain microvasculature, astrocytes, and red blood cells. Most *SLC2A1* mutations reported to date appear to manifest as haploinsufficiency. It is possible that paroxysmal movement disorders are caused by astrocyte failure in critical regions of motor system such as the cerebellum (Schneider *et al.*, 2009).

3. DYT10/DYT19 (PKD)

Using linkage and haplotype analysis, Valente *et al.* (2000) identified a 15.8-cm candidate region for PKD in a large Indian family. The candidate region is flanked by markers D16S685 and D16S503 on Chr 16q13-q22.1 with a maximum LOD score of 3.66 at D16S419. This candidate region is telomeric to the locus identified in Japanese families with PKD (Tomita *et al.*, 1999) but showed overlap with a region identified in an African-American family with PKD (Bennett *et al.*, 2000). The candidate region for infantile familial convulsions and paroxysmal choreoathetosis (ICCA) was also mapped to the pericentromeric region of Chr 16 in both French (Szepetowski *et al.*, 1997) and Chinese (Lee *et al.*, 1998) families. PKD is genetically heterogeneous, however, and, in at least one British pedigree, does not map to Chr 16 (Spacey *et al.*, 2002).

The family described by Valente *et al.* (2000) included 13 subjects with PKD. In addition, four deceased family members had probably been affected based on historical accounts. Age at onset ranged from 7 to 13 years. Consistent with PKD, attacks were precipitated by sudden movements and lasted less than 2 min. Attacks included choreiform and dystonic movements and occurred up to 20 times daily. Two family members with PKD and three unaffected family members had generalized tonic–clonic seizures as teenagers. However, epilepsy did not co-segregate with the PKD haplotype. Penetrance was 75%, and over half of the affected family members had spontaneous remissions in their early twenties (Spacey *et al.*, 2002).

Just recently, several distinct loss-of-function frameshift mutations leading to protein truncation or nonsense-mediated decay in proline-rich transmembrane protein 2 (*PRRT2*) have been associated with PKD in numerous Han Chinese families (Chen *et al.*, 2011; Li *et al.*, 2012; Liu *et al.*, 2012; Wang *et al.*, 2011). Missense mutations (c.796C>T, p.R266W; c.913G>A, p.G305R) have been identified in a single family and one sporadic case of PKD (Liu *et al.*, 2012; Wang *et al.*, 2011). PRRT2 contains two predicted transmembrane domains and is highly expressed in the developing nervous system, particularly the cerebellum (Chen *et al.*, 2011). *PRRT2* is located on Chr 16p.11.2, within the ICCA, Japanese PKD, and African-American candidate regions but outside the Indian PKD candidate regions. In addition to classic carbamazepine-responsive PKD, the phenotypic spectrum of PRRT2 mutations appears to include ICCA, a "PNKD-like" syndrome and PED (Liu *et al.*, 2012).

4. DYT20 (PNKD2)

PNKD is genetically heterogeneous (Bruno *et al.*, 2007). For instance, Spacey *et al.* (2006) described a Canadian family with PNKD and without a mutation in *PNKD*. In this kindred, the PNKD2 locus maps to Chr 2p31.

B. Ataxias

Dystonia can be a prominent clinical feature or presenting sign in many of the dominant (van Gaalen *et al.*, 2011) and recessive (Perlman, 2011) ataxias. For example, leg or hand–forearm dystonia can be presenting signs in spinocerebellar ataxia type 6 (SCA6) and SCA17 (Hagenah *et al.*, 2004; Sethi and Jankovic, 2002). Similarly, cervical dystonia has been reported as a presenting sign in SCA1 (Wu *et al.*, 2004) and SCA6 (Arpa *et al.*, 1999). Among the SCAs, dystonia is most common in SCA17, SCA3, and SCA2 (van Gaalen *et al.*, 2011).

Although relatively infrequent, dystonia, appendicular and cervical, has been well documented in Friedrich ataxia (FA) (Hou and Jankovic, 2003). FA is perhaps the most common and intensively studied of the hereditary ataxias. FA is caused by reduced expression of the mitochondrial protein frataxin due to a GAA trinucleotide expansion within Intron 1 of *FXN*. Frataxin plays a critical role in mitochondrial iron metabolism (Shan and Cortopassi, 2011), and FA can be considered a mitochondrial disease caused by a mutation in nuclear DNA (nDNA). In some patients, point mutations are present in heterozygosity. Dystonia may also manifest in several other recessive and X-linked ataxias (Table 2.3).

Le Ber *et al.* (2006) reported eight families characterized by the combination of dystonia and MRI evidence of cerebellar atrophy. Age of dystonia onset ranged from 9 to 42 years. Most individuals had clinical evidence of laryngeal or arm dystonia at disease onset. Dystonia generalized in 5/12 individuals. Cerebellar ataxia was relatively mild and slowly progressive.

Table 2.3 Dystonia associated with other neurogenetic disorders

Group	Disorder	MIM	Gene	Inheritance	Protein	Phenotype
Dominant ataxias	SCA1	164400	*ATXN1*	AD	Ataxin-1	Ataxia, slow saccades, dysarthria, corticospinal signs, dystonia
	SCA2	183090	*ATXN2*	AD	Ataxin-2	Ataxia, slow saccades, dysarthria, hyporeflexia, dystonia
	SCA3	109150	*ATXN3*	AD	Ataxin-3	Ataxia, ophthalmoplegia, parkinsonism, dystonia, spasticity, neuropathy
	SCA6	183086	*CACNA1A*	AD	α-1A subunit P/Q voltage-gated Ca^{2+} channel	Ataxia, nystagmus, dystonia
	SCA7	164500	*ATXN7*	AD	Ataxin-7	Ataxia, macular degeneration, dystonia, spasticity
	SCA8	608768	*ATXN8*	AD	Ataxin-8	Ataxia, nystagmus, spasticity, dystonia, sensory neuropathy
	SCA12	604326	*PPP2R2B*	AD	Serine/threonine-protein phosphatase 2A 55-kDa regulatory subunit B, β isoform	Ataxia, facial myokymia, action tremor, dystonia, parkinsonism, dementia
	SCA14	605361	*PRKCG*	AD	Protein kinase C, gamma	Ataxia, myoclonus, slow saccades, focal dystonia
	SCA17	607136	*TBP*	AD	TATA box-binding protein	Ataxia, spasticity, seizures, dystonia, dementia
	Dentatorubral-pallidoluysian atrophy (DRPLA)	125370	*ATN1*	AD	Atrophin 1	Ataxia, myoclonus, seizures, dystonia

Recessive ataxias	Friedreich ataxia	606829	*FXN*	AR	Frataxin	Ataxia, nystagmus, optic atrophy, cardiomyopathy, sensory neuropathy, spasticity, dystonia, diabetes, scoliosis, pes cavus
	Ataxia telangiectasia	208900	*AT*	AR	Serine-protein kinase ATM	Ataxia, telangiectasis, oculomotor apraxia, immune deficiency, seizures, dystonia
	Ataxia with oculomotor apraxia type 1	208920	*APTX*	AR	Aprataxin	Ataxia, oculomotor apraxia, scoliosis, pes cavus, dystonia, choreoathetosis, neuropathy
	Ataxia with oculomotor apraxia type 2	606002	*SETX*	AR	Senataxin	Ataxia, oculomotor apraxia, nystagmus, amyotrophy, dystonia, chorea, neuropathy
	Ataxia with selective vitamin E deficiency	277460	*TTPA*	AR	α-Tocopherol transfer protein	Ataxia, areflexia, loss of proprioception, dystonia, xanthelasma, tendon xanthomas
X-linked ataxias	Mental retardation, X-linked syndromic, Christianson type	300243	*SLC9A6*	X-linked dominant	Sodium/hydrogen exchanger 6	Ataxia, spasticity, mental retardation, dysmorphic features, microcephaly, dystonia, autistic
Dopamine metabolism	AADC deficiency	608643	*AADC*	AR	Aromatic L-amino acid decarboxylase	Psychomotor delay, ptosis, oculogyric crises, dystonia, truncal hypotonia, temperature instability, hypotension
	Infantile parkinsonism–dystonia	613135	*SLC6A3*	AR	Presynaptic dopamine transporter	Parkinsonism, dystonia, developmental delay, upper motor neuron signs

(*Continued*)

Table 2.3 Dystonia associated with other neurogenetic disorders—cont'd

Group	Disorder	MIM	Gene	Inheritance	Protein	Phenotype
Deafness	Mohr–Tranebjaerg syndrome	300356	*TIMM8A*	X-linked	Translocase of inner mitochondrial membrane 8	Sensorineural deafness at an early age followed by varying degrees of dystonia along with visual and cognitive disability
	Woodhouse–Sakati syndrome	241080	*DCAF17*	AR	Nucleolar protein	Sensorineural deafness, seizures, sensory neuropathy, mental retardation, alopecia, hypogonadism, diabetes mellitus
Spasticity	Spastic paraplegia 35 (SPG35)	612319	*FA2H*	AR	Fatty acid 2-hydroxylase	Spasticity, cognitive decline, dystonia, optic atrophy, brain iron accumulation, leukodystrophy
	SPG2, Pelizaeus–Merzbacher disease	312920, 312080	*PLP1*	X-linked	Proteolipid protein-1	Spasticity, ataxia, dystonia, developmental delay, hypomyelinative leukodystrophy
	SPG15	ZFYVE26	*KI*	AR	Spastizin	Spasticity, amyotrophy, ataxia, mental retardation, dystonia
Metabolic and storage	Wilson disease	277900	*ATP7B*	AR	ATPase, Cu^{2+} transporting β polypeptide	Hepatitis, liver failure, Kayser–Fleischer rings, tremor, dystonia, personality changes
	Lesch–Nyhan syndrome	300322	*HPRT1*	X-linked recessive	Hypoxanthine guanine phosphoribosyl-transferase I	Hyperuricemia, gout, nephrolithiasis, self-injurious behavior, dystonia, mental retardation, athetosis

Glutaric academia I	231670	*GCDH*	AR	Glutaryl-CoA dehydrogenase	Macrocephaly, hepatomegaly, dystonia, choreoathetosis, infantile encephalopathy, striatal necrosis
Pyruvate dehydrogenase deficiency	312170	*PDHA1*	X-linked dominant	Pyruvate dehydrogenase E1-α deficiency	Low birth weight, microcephaly, dysmorphic features, psychomotor retardation, seizures, choreoathetosis, dystonia, ataxia, lactic acidosis
Homocystinuria	236200	*CBS*	AR	Cystathionine β-synthase deficiency	Tall stature, ectopic lentis, skeletal abnormalities, seizures, mental retardation, dystonia, thromboembolism
Biotin-responsive basal ganglia disease	607483		AR		Subacute encephalopathy, dysphagia, dysarthria, dystonia, nystagmus, seizures
Niemann—Pick type C1	257220	*NPC1*	AR	Neimann-Pick C1 protein	Hepatosplenomegaly, supranuclear gaze palsy, dementia, dysarthria, dystonia, ataxia
Niemann—Pick type C2	607625	*NPC2*	AR	Epididymal secretory protein E1	Hepatosplenomegaly, respiratory failure, supranuclear gaze play, dementia, dysarthria, dystonia, ataxia
GM1 gangliosidosis	230500	*GLB1*	AR	β-Galactosidase	Dysmorphic face/neck, dwarfism, macular cherry-red spot, hepatosplenomegaly, mental retardation, dystonia

(*Continued*)

Table 2.3 Dystonia associated with other neurogenetic disorders—cont'd

Group	Disorder	MIM	Gene	Inheritance	Protein	Phenotype
	GM2 gangliosidosis	272750	*GM2A*	AR	Ganglioside GM2 activator	Macular cherry-red spot, blindness, psychomotor delay, dementia, seizures, dystonia
	Tay—Sachs disease	272800	*HEXA*	AR	β-Hexosaminidase, subunit α	Macular cherry-red spot, blindness, psychomotor deterioration, seizures, dementia, dystonia
	Neuronal ceroid-lipofuscinosis[a]	204200	*CLN3*	AR	Battenin	Retinitis pigmentosa with progressive vision loss, psychomotor degeneration with dementia, parkinsonism, ataxia, dystonia, seizures
	Metachromatic leukodystrophy	250100	*ARSA*	AR	Arylsulfatase A	Optic atrophy, dementia, ataxia, chorea, dystonia, seizures, demyelinating polyneuropathy
Mitochondrial	Leber hereditary optic atrophy (LHON)	53500	*MTND1*[b]	M	NADH dehydrogenase I	Optic atrophy, action tremor, dystonia, peripheral neuropathy
	Leigh syndrome[c]	25600	*MTND2*	AR, M, X-linked	NAD dehydrogenase subunit 2	Ophthalmoplegia, pigmentary retinopathy, psychomotor retardation, ataxia, dystonia, spasticity, seizures, lactic acidosis
Other neuro-degenerative disorders	Rett syndrome	312750	*MECP2*	X-linked dominant	MECP2	Mental retardation, stereotypies, parkinsonism, dystonia
	Huntington disease (HD)	143100	*HTT*	AD	Huntington	Chorea, personality changes, dementia, dystonia

Huntington disease-like 2 (HDL2)	606438	*JPH3*	AD	Junctophilin 3	Dementia, akinetic-rigid syndrome, mild chorea, mild dystonia
Choreoacanth-ocytosis	200150	*VPS13A*	AR	Chorein	Orofacial dyskinesias, dysphagia, choreoathetosis, dystonia, seizures, parkinsonism, acanthocytosis
Creutzfeldt–Jakob disease, familial	123400	*PRNP*	AD	Prion protein	Dementia, delirium, myoclonus, akinetic-rigid syndrome, dystonia, ataxia

AD, autosomal dominant; AR, autosomal recessive; M, mitochondrial.
[a]Neuronal ceroid lipofuscinoses are a group of genetically heterogeneous disorders characterized by intracellular accumulation of autofluorescent lipopigment and include dementia, movement disorders, vision loss, and seizures as phenotypic features (Getty and Pearce, 2011).
[b]LHON has been associated with variants in *ND1*, *ND2*, *ND4*, *ND4L*, *ND5*, *ND6*, *MTCYB*, *MTCO1*, and *MTAP6* (Tońska *et al.*, 2010).
[c]Leigh syndrome is associated with marked genetic hetereogeneity and can be caused by mutations in either mtDNA or nDNA (Finsterer, 2008).

Inheritance was either recessive or X-linked. Similar families have been described by other groups (Hagenah *et al.*, 2007).

C. Other

Dystonia may be a clinical sign in a diverse array of neurogenetic disorders including those associated with autism and mental retardation, mitochondrial dysfunction, deafness, spasticity, dopamine neurotransmission, and metabolic storage defects (Table 2.3). In many of these conditions, the functional impact of dystonia on quality of life is relatively minor compared to the effects of the causal mutation on cognition, hearing, fine and gross motor skills, or extra-neural systems. There are several exceptions, however, and dystonia can occasionally be a presenting feature or major source of disability.

Dystonia-deafness syndrome (Mohr–Tranebjaerg syndrome) is an X-linked recessive disorder caused by mutations in *DDP1*, which encodes deafness-dystonia peptide 1 or translocase of inner mitochondrial membrane 8. Males develop childhood-onset deafness, spasticity, blindness, and dystonia. Dystonia can generalize and affect the craniocervical musculature (Kim *et al.*, 2007). Cervical dystonia and hand–forearm dystonia (e.g., writer's cramp) have been reported in female carriers (Swerdlow and Wooten, 2001). The combination of deafness and dystonia is also seen in the Woodhouse–Sakati syndrome (Schneider and Bhatia, 2008), which is due to mutations in *DCAF17* that encodes a nucleolar protein (Alazami *et al.*, 2008).

Spasticity and dystonia may coexist as "spastic dystonia" or "spastic ataxia" syndromes or in a subset of patients with spastic paraplegia 35 (SPG35 or FAHN disease) and the allelic disorders, SGP2 and Pelizaeus–Merzbacher disease (Table 2.3). Gilbert *et al.* (2009) have mapped a novel spastic dystonia locus to Chr 2q24-31. Their Ohio kindred manifests spasticity and dystonia without ataxia or dementia. One subject showed marked improvement of dystonia with deep brain stimulation. Portneuf spastic ataxia with eukoencephalopathy is an autosomal recessive disorder mapped to 2q33-34 and associated with spasticity, ataxia, mild cognitive impairment, dystonia, nystagmus, and scoliosis (Thiffault *et al.*, 2006).

Screening ceruloplasmin and uric acid levels should be obtained in all children and young adults presenting with dystonia to exclude Wilson and Lesch–Nyhan diseases, respectively. Wilson disease is due to recessive mutations in *ATP7B*, which encodes an ATPase, Cu^{2+} transporting β polypeptide. In developed countries, Wilson disease is often diagnosed prior to the onset of significant neurological dysfunction. Virtually all patients with neurological disease have Kayser–Fleischer rings due to copper deposition in Descemet membrane. Dystonia is often among the protean neuropsychiatric manifestations of more advanced Wilson disease. Severe generalized dystonia, mental retardation, gout, and self-injurious behavior are seen is the classic form of Lesch–Nyhan disease. Attenuated variants of this disorder may present with isolated focal or segmental dystonia (Jinnah *et al.*, 2010). Nonsense and deletion mutations are usually associated with classic Lesch–Nyhan disease (Jinnah *et al.*, 2010).

Dystonia can be a manifestation of disorders linked to mitochondrial dysfunction due to mutations in either nDNA or mtDNA. Examples of dystonia-associated nuclear genes include *TIMM8A* in X-linked dystonia-deafness syndrome and *TACO1* in Leigh syndrome (Weraarpachai *et al.*, 2009). An enormous array of mtDNA mutations have been tied to Leber hereditary optic neuropathy (Tonska *et al.*, 2010) and Leigh syndrome (Finsterer, 2008). Mitochondrial dysfunction may also be a feature of more common forms of dystonia. For example, Schapira *et al.* (1997) found a 22% decrease in the activity of complex I, but not complex II/III or complex IV, in 20 patients with idiopathic focal dystonia compared to 22 controls. Similarly, Benecke *et al.* (1992) found a 37% reduction in complex I activity in patients with focal dystonia and a 67% decrease in complex I activity in eight patients with segmental or generalized dystonia. These data suggest that sequence variants in mtDNA may contribute to risk for largely sporadic adult-onset primary dystonia.

Dystonia is an important clinical feature in several of the hereditary and neurodegenerative choreas, particularly Huntington disease (*HTT*) and choreoacanthocystosis (*VPS13A*). The combination of chorea and dystonia may also been seen in familial or idiopathic basal ganglia calcifications (IBGC). In a large family with autosomal dominant IBGC, the disease locus was mapped to 14q (Geschwind *et al.*, 1999). Many other families with IBGC do not show linkage to 14q (Oliveira *et al.*, 2004).

VII. CONCLUSIONS

Major recent discoveries have appeared in the field of dystonia genetics. First, mutations in the transcription factor THAP1 and DNA replication factor CIZ1 have been linked to adult-onset primary dystonia (Xiao *et al.*, 2010, 2012). Second, proline-rich transmembrane protein 2 has been tied to paroxysmal kinesigenic dyskinesia (Chen *et al.*, 2011). Third, there has been consolidation of DYT syndromes—DYT9 (paroxysmal choreoathetosis/spasticity) and DYT18 (paroxysmal exertional dyskinesia) are now known to be treatable allelic disorders, and DYT14 has been redefined as DYT5 due to a deletion mutation in *GCH1* (Wider *et al.*, 2008).

Recent advances in high-throughput sequencing and bioinformatics should enable additional discoveries. A major focus of future efforts should be the genetics of adult-onset primary dystonia. Unfortunately, this work will be compromised by the paucity of large kindreds with highly penetrant disease. Although there is no good evidence that adult-onset primary dystonia is a complex disease, well-designed genome-wide association studies of specific forms of focal dystonia may point out loci that contribute to disease penetrance. Work to date has implicated several cellular pathways in the pathophysiology of dystonia. These include G1/S cell cycle control in primary dystonia; and monoamine neurotransmission, the neuronal stress response, and mitochondrial dysfunction in non-primary dystonia. Hopefully, future work can serve to unify, in whole or part, the cellular and systems biology of dystonia.

ACKNOWLEDGMENTS

My work on dystonia has been supported by the Neuroscience Institute at the University of Tennessee Health Science Center, Dystonia Medical Research Foundation, NIH grants R01NS048458 and R01NS069936, NIH U54 Dystonia Coalition (1U54NS065701) Pilot Projects Program, and the Parkinson's & Movement Disorder Foundation.

REFERENCES

Alazami, A.M., Al-Saif, A., Al-Semari, A., Bohlega, S., Zlitni, S., Alzahrani, F., Bavi, P., Kaya, N., Colak, D., Khalak, H., et al., 2008. Mutations in C2orf37, encoding a nucleolar protein, cause hypogonadism, alopecia, diabetes mellitus, mental retardation, and extrapyramidal syndrome. Am. J. Hum. Genet. 83, 684–691.

Aquino, D., Bizzi, A., Grisoli, M., Garavaglia, B., Bruzzone, M.G., Nardocci, N., Savoiardo, M., Chiapparini, L., 2009. Age-related iron deposition in the basal ganglia: quantitative analysis in healthy subjects. Radiology 252, 165–172.

Arpa, J., Cuesta, A., Cruz-Martinez, A., Santiago, S., Sarria, J., Palau, F., 1999. Clinical features and genetic analysis of a Spanish family with spinocerebellar ataxia 6. Acta Neurol. Scand. 99, 43–47.

Asmus, F., Salih, F., Hjermind, L.E., Ostergaard, K., Munz, M., Kuhn, A.A., Dupont, E., Kupsch, A., Gasser, T., 2005. Myoclonus-dystonia due to genomic deletions in the epsilon-sarcoglycan gene. Ann. Neurol. 58, 792–797.

Auburger, G., Ratzlaff, T., Lunkes, A., Nelles, H.W., Leube, B., Binkofski, F., Kugel, H., Heindel, W., Seitz, R., Benecke, R., et al., 1996. A gene for autosomal dominant paroxysmal choreoathetosis/spasticity (CSE) maps to the vicinity of a potassium channel gene cluster on chromosome 1p, probably within 2 cM between D1S443 and D1S197. Genomics 31, 90–94.

Augood, S.J., Keller-McGandy, C.E., Siriani, A., Hewett, J., Ramesh, V., Sapp, E., DiFiglia, M., Breakefield, X.O., Standaert, D.G., 2003. Distribution and ultrastructural localization of torsinA immunoreactivity in the human brain. Brain Res. 986, 12–21.
Baker, M., Litvan, I., Houlden, H., Adamson, J., Dickson, D., Perez-Tur, J., Hardy, J., Lynch, T., Bigio, E., Hutton, M., 1999. Association of an extended haplotype in the tau gene with progressive supranuclear palsy. Hum. Mol. Genet. 8, 711–715.
Barclay, C.L., Lang, A.E., 1997. Dystonia in progressive supranuclear palsy. J. Neurol. Neurosurg. Psychiatry 62, 352–356.
Basham, S.E., Rose, L.S., 2001. The Caenorhabditis elegans polarity gene *ooc-5* encodes a Torsin-related protein of the AAA ATPase superfamily. Development 128, 4645–4656.
Benecke, R., Strumper, P., Weiss, H., 1992. Electron transfer complex I defect in idiopathic dystonia. Ann. Neurol. 32, 683–686.
Bennett, L.B., Roach, E.S., Bowcock, A.M., 2000. A locus for paroxysmal kinesigenic dyskinesia maps to human chromosome 16. Neurology 54, 125–130.
Bentivoglio, A.R., Ialongo, T., Contarino, M.F., Valente, E.M., Albanese, A., 2004. Phenotypic characterization of DYT13 primary torsion dystonia. Mov. Disord. 19, 200–206.
Bhatia, K.P., 2011. Paroxysmal dyskinesias. Mov. Disord. 26, 1157–1165.
Bonafé, L., Thöny, B., Penzien, J.M., Czarnecki, B., Blau, N., 2001. Mutations in the sepiapterin reductase gene cause a novel tetrahydrobiopterin-dependent monoamine-neurotransmitter deficiency without hyperphenylalaninemia. Am. J. Hum. Genet. 69, 269–277.
Brashear, A., Dobyns, W.B., de Carvalho Aguiar, P., Borg, M., Frijns, C.J., Gollamudi, S., Green, A., Guimaraes, J., Haake, B.C., Klein, C., et al., 2007. The phenotypic spectrum of rapid-onset dystonia-parkinsonism (RDP) and mutations in the ATP1A3 gene. Brain 130, 828–835.
Braun, N., Zhu, Y., Krieglstein, J., Culmsee, C., Zimmermann, H., 1998. Upregulation of the enzyme chain hydrolyzing extracellular ATP after transient forebrain ischemia in the rat. J. Neurosci. 18, 4891–4900.
Bressman, S.B., Raymond, D., Fuchs, T., Heiman, G.A., Ozelius, L.J., Saunders-Pullman, R., 2009. Mutations in THAP1 (DYT6) in early-onset dystonia: a genetic screening study. Lancet Neurol. 8, 441–446.
Bressman, S.B., Sabatti, C., Raymond, D., de Leon, D., Klein, C., Kramer, P.L., Brin, M.F., Fahn, S., Breakefield, X., Ozelius, L.J., et al., 2000. The DYT1 phenotype and guidelines for diagnostic testing. Neurology 54, 1746–1752.
Bruno, M.K., Lee, H.Y., Auburger, G.W., Friedman, A., Nielsen, J.E., Lang, A.E., Bertini, E., Van Bogaert, P., Averyanov, Y., Hallett, M., et al., 2007. Genotype-phenotype correlation of paroxysmal nonkinesigenic dyskinesia. Neurology 68, 1782–1789.
Calakos, N., Patel, V.D., Gottron, M., Wang, G., Tran-Viet, K.N., Brewington, D., Beyer, J.L., Steffens, D.C., Krishnan, R.R., Zuchner, S., 2010. Functional evidence implicating a novel TOR1A mutation in idiopathic, late-onset focal dystonia. J. Med. Genetic. 47, 646–650.
Caldwell, G.A., Cao, S., Sexton, E.G., Gelwix, C.C., Bevel, J.P., Caldwell, K.A., 2003. Suppression of polyglutamine-induced protein aggregation in Caenorhabditis elegans by torsin proteins. Hum. Mol. Genet. 12, 307–319.
Callan, A.C., Bunning, S., Jones, O.T., High, S., Swanton, E., 2007. Biosynthesis of the dystonia-associated AAA+ ATPase torsinA at the endoplasmic reticulum. Biochem. J. 401, 607–612.
Camargos, S., Scholz, S., Simon-Sanchez, J., Paisan-Ruiz, C., Lewis, P., Hernandez, D., Ding, J., Gibbs, J.R., Cookson, M.R., Bras, J., et al., 2008. DYT16, a novel young-onset dystonia-parkinsonism disorder: identification of a segregating mutation in the stress-response protein PRKRA. Lancet Neurol. 7, 207–215.

Cao, S., Gelwix, C.C., Caldwell, K.A., Caldwell, G.A., 2005. Torsin-mediated protection from cellular stress in the dopaminergic neurons of Caenorhabditis elegans. J. Neurosci. 25, 3801–3812.
Chen, W.J., Lin, Y., Xiong, Z.Q., Wei, W., Ni, W., Tan, G.H., Guo, S.L., He, J., Chen, Y.F., Zhang, Q.J., et al., 2011. Exome sequencing identifies truncating mutations in PRRT2 that cause paroxysmal kinesigenic dyskinesia. Nat. Genet. 43, 1252–1255.
Cheng, F.B., Ozelius, L.J., Wan, X.H., Feng, J.C., Ma, L.Y., Yang, Y.M., Wang, L., 2011. THAP1/DYT6 sequence variants in non-DYT1 early-onset primary dystonia in China and their effects on RNA expression. J. Neurol. 259, 342–347.
Chouery, E., Kfoury, J., Delague, V., Jalkh, N., Bejjani, P., Serre, J.L., Megarbane, A., 2008. A novel locus for autosomal recessive primary torsion dystonia (DYT17) maps to 20p11.22-q13.12. Neurogenetics 9, 287–293.
Choumert, A., Poisson, A., Honnorat, J., Le Ber, I., Camuzat, A., Broussolle, E., Thobois, S., 2011. G303V tau mutation presenting with progressive supranuclear palsy-like features. Mov. Disord. http://dx.doi.org/10.1002/mds.24060.
Clouaire, T., Roussigne, M., Ecochard, V., Mathe, C., Amalric, F., Girard, J.P., 2005. The THAP domain of THAP1 is a large C2CH module with zinc-dependent sequence-specific DNA-binding activity. Proc. Natl. Acad. Sci. U. S. A. 102, 6907–6912.
de Carvalho Aguiar, P., Sweadner, K.J., Penniston, J.T., Zaremba, J., Liu, L., Caton, M., Linazasoro, G., Borg, M., Tijssen, M.A., Bressman, S.B., et al., 2004. Mutations in the Na+/K+ -ATPase alpha3 gene ATP1A3 are associated with rapid-onset dystonia parkinsonism. Neuron 43, 169–175.
Defazio, G., Abbruzzese, G., Aniello, M.S., Bloise, M., Crisci, C., Eleopra, R., Fabbrini, G., Girlanda, P., Liguori, R., Macerollo, A., et al., 2011. Environmental risk factors and clinical phenotype in familial and sporadic primary blepharospasm. Neurology 77, 631–637.
Djarmati, A., Schneider, S.A., Lohmann, K., Winkler, S., Pawlack, H., Hagenah, J., Bruggemann, N., Zittel, S., Fuchs, T., Rakovic, A., et al., 2009. Mutations in THAP1 (DYT6) and generalised dystonia with prominent spasmodic dysphonia: a genetic screening study. Lancet Neurol. 8, 447–452.
Dudesek, A., Roschinger, W., Muntau, A.C., Seidel, J., Leupold, D., Thöny, B., Blau, N., 2001. Molecular analysis and long-term follow-up of patients with different forms of 6-pyruvoyl-tetrahydropterin synthase deficiency. Eur. J. Pediatr. 160, 267–276.
Edwards, M., Wood, N., Bhatia, K., 2003. Unusual phenotypes in DYT1 dystonia: a report of five cases and a review of the literature. Mov. Disord. 18, 706–711.
Einholm, A.P., Toustrup-Jensen, M.S., Holm, R., Andersen, J.P., Vilsen, B., 2010. The rapid-onset dystonia parkinsonism mutation D923N of the Na+, K+-ATPase alpha3 isoform disrupts Na+ interaction at the third Na+ site. J. Biol. Chem. 285, 26245–26254.
Esapa, C.T., Waite, A., Locke, M., Benson, M.A., Kraus, M., McIlhinney, R.A., Sillitoe, R.V., Beesley, P.W., Blake, D.J., 2007. SGCE missense mutations that cause myoclonus-dystonia syndrome impair epsilon-sarcoglycan trafficking to the plasma membrane: modulation by ubiquitination and torsinA. Hum. Mol. Genet. 16, 327–342.
Evidente, V.G., Advincula, J., Esteban, R., Pasco, P., Alfon, J.A., Natividad, F.F., Cuanang, J., Luis, A.S., Gwinn-Hardy, K., Hardy, J., et al., 2002. Phenomenology of "Lubag" or X-linked dystonia-parkinsonism. Mov. Disord. 17, 1271–1277.
Fahn, S., 1987. Systemic therapy of dystonia. Can. J. Neurol. Sci. 14, 528–532.
Fahn, S., 1988. Concept and classification of dystonia. Adv. Neurol. 50, 1–8.
Fahn, S., 2011. Classification of movement disorders. Mov. Disord. 26, 947–957.
Fahn, S., Bressman, S.B., Marsden, C.D., 1998. Classification of dystonia. Adv. Neurol. 78, 1–10.

Ferrari-Toninelli, G., Paccioretti, S., Francisconi, S., Uberti, D., Memo, M., 2004. TorsinA negatively controls neurite outgrowth of SH-SY5Y human neuronal cell line. Brain Res. 1012, 75–81.
Finsterer, J., 2008. Leigh and Leigh-like syndrome in children and adults. Pediatr. Neurol. 39, 223–235.
Frédéric, M., Lucarz, E., Monino, C., Saquet, C., Thorel, D., Claustres, M., Tuffery-Giraud, S., Collod-Beroud, G., 2007. First determination of the incidence of the unique TOR1A gene mutation, c.907delGAG, in a Mediterranean population. Mov. Disord. 22, 884–888.
Frédéric, M.Y., Clot, F., Blanchard, A., Dhaenens, C.M., Lesca, G., Cif, L., Durr, A., Vidailhet, M., Sablonniere, B., Calender, A., et al., 2009. The p.Asp216His TOR1A allele effect is not found in the French population. Mov. Disord. 24, 919–921.
Fuchs, T., Gavarini, S., Saunders-Pullman, R., Raymond, D., Ehrlich, M.E., Bressman, S.B., Ozelius, L.J., 2009. Mutations in the THAP1 gene are responsible for DYT6 primary torsion dystonia. Nat. Genet. 41, 286–288.
Gajos, A., Piaskowski, S., Slawek, J., Ochudlo, S., Opala, G., Lobinska, A., Honczarenko, K., Budrewicz, S., Koszewicz, M., Pelszynska, B., et al., 2007. Phenotype of the DYT1 mutation in the TOR1A gene in a Polish population of patients with dystonia. A preliminary report. Neurol. Neurochir. Pol. 41, 487–494.
Gambarin, M., Valente, E.M., Liberini, P., Barrano, G., Bonizzato, A., Padovani, A., Moretto, G., Fiorio, M., Dallapiccola, B., Smania, N., et al., 2006. Atypical phenotypes and clinical variability in a large Italian family with DYT1-primary torsion dystonia. Mov. Disord. 21, 1782–1784.
Gasser, T., Windgassen, K., Bereznai, B., Kabus, C., Ludolph, A.C., 1998. Phenotypic expression of the DYT1 mutation: a family with writer's cramp of juvenile onset. Ann. Neurol. 44, 126–128.
Geschwind, D.H., Loginov, M., Stern, J.M., 1999. Identification of a locus on chromosome 14q for idiopathic basal ganglia calcification (Fahr disease). Am. J. Hum. Genet. 65, 764–772.
Getty, A.L., Pearce, D.A., 2011. Interactions of the proteins of neuronal ceroid lipofuscinosis: clues to function. Cell. Mol. Life Sci. 68, 453–474.
Ghezzi, D., Viscomi, C., Ferlini, A., Gualandi, F., Mereghetti, P., DeGrandis, D., Zeviani, M., 2009. Paroxysmal non-kinesigenic dyskinesia is caused by mutations of the MR-1 mitochondrial targeting sequence. Hum. Mol. Genet. 18, 1058–1064.
Gilbert, D.L., Leslie, E.J., Keddache, M., Leslie, N.D., 2009. A novel hereditary spastic paraplegia with dystonia linked to chromosome 2q24-2q31. Mov. Disord. 24, 364–370.
Gimenez-Roldan, S., Delgado, G., Marin, M., Villanueva, J.A., Mateo, D., 1988. Hereditary torsion dystonia in gypsies. Adv. Neurol. 50, 73–81.
Goetz, C.G., Chmura, T.A., Lanska, D.J., 2001. History of dystonia: part 4 of the MDS-sponsored history of movement disorders exhibit, Barcelona, June, 2000. Mov. Disord. 16, 339–345.
Goodchild, R.E., Dauer, W.T., 2004. Mislocalization to the nuclear envelope: an effect of the dystonia-causing torsinA mutation. Proc. Natl. Acad. Sci. U. S. A. 101, 847–852.
Goodchild, R.E., Dauer, W.T., 2005. The AAA+ protein torsinA interacts with a conserved domain present in LAP1 and a novel ER protein. J. Cell Biol. 168, 855–862.
Goodchild, R.E., Kim, C.E., Dauer, W.T., 2005. Loss of the dystonia-associated protein torsinA selectively disrupts the neuronal nuclear envelope. Neuron 48, 923–932.
Goto, S., Lee, L.V., Munoz, E.L., Tooyama, I., Tamiya, G., Makino, S., Ando, S., Dantes, M.B., Yamada, K., Matsumoto, S., et al., 2005. Functional anatomy of the basal ganglia in X-linked recessive dystonia-parkinsonism. Ann. Neurol. 58, 7–17.
Groen, J.L., Ritz, K., Contarino, M.F., van de Warrenburg, B.P., Aramideh, M., Foncke, E.M., van Hilten, J.J., Schuurman, P.R., Speelman, J.D., Koelman, J.H., et al.,

2010. DYT6 dystonia: mutation screening, phenotype, and response to deep brain stimulation. Mov. Disord. 25, 2420–2427.
Grotzsch, H., Pizzolato, G.P., Ghika, J., Schorderet, D., Vingerhoets, F.J., Landis, T., Burkhard, P.R., 2002. Neuropathology of a case of dopa-responsive dystonia associated with a new genetic locus, DYT14. Neurology 58, 1839–1842.
Grundmann, K., Laubis-Herrmann, U., Bauer, I., Dressler, D., Vollmer-Haase, J., Bauer, P., Stuhrmann, M., Schulte, T., Schols, L., Topka, H., et al., 2003. Frequency and phenotypic variability of the GAG deletion of the DYT1 gene in an unselected group of patients with dystonia. Arch. Neurol. 60, 1266–1270.
Grunewald, A., Djarmati, A., Lohmann-Hedrich, K., Farrell, K., Zeller, J.A., Allert, N., Papengut, F., Petersen, B., Fung, V., Sue, C.M., et al., 2008. Myoclonus-dystonia: significance of large SGCE deletions. Hum. Mutat. 29, 331–332.
Hagenah, J., Reetz, K., Zuhlke, C., Rolfs, A., Binkofski, F., Klein, C., 2007. Predominant dystonia with marked cerebellar atrophy: a rare phenotype in familial dystonia. Neurology 68, 2157.
Hagenah, J.M., Zuhlke, C., Hellenbroich, Y., Heide, W., Klein, C., 2004. Focal dystonia as a presenting sign of spinocerebellar ataxia 17. Mov. Disord. 19, 217–220.
Hedera, P., Phibbs, F.T., Fang, J.Y., Cooper, M.K., Charles, P.D., Davis, T.L., 2010. Clustering of dystonia in some pedigrees with autosomal dominant essential tremor suggests the existence of a distinct subtype of essential tremor. BMC Neurol. 10, 66.
Hewett, J.W., Zeng, J., Niland, B.P., Bragg, D.C., Breakefield, X.O., 2006. Dystonia-causing mutant torsinA inhibits cell adhesion and neurite extension through interference with cytoskeletal dynamics. Neurobiol. Dis. 22, 98–111.
Holmgren, G., Ozelius, L., Forsgren, L., Almay, B.G., Holmberg, M., Kramer, P., Fahn, S., Breakefield, X.O., 1995. Adult onset idiopathic torsion dystonia is excluded from the DYT 1 region (9q34) in a Swedish family. J. Neurol. Neurosurg. Psychiatry 59, 178–181.
Horvath, G.A., Stockler-Ipsiroglu, S.G., Salvarinova-Zivkovic, R., Lillquist, Y.P., Connolly, M., Hyland, K., Blau, N., Rupar, T., Waters, P.J., 2008. Autosomal recessive GTP cyclohydrolase I deficiency without hyperphenylalaninemia: evidence of a phenotypic continuum between dominant and recessive forms. Mol. Genet. Metab. 94, 127–131.
Hou, J.G., Jankovic, J., 2003. Movement disorders in Friedreich's ataxia. J. Neurol. Sci. 206, 59–64.
Houlden, H., Baker, M., Morris, H.R., MacDonald, N., Pickering-Brown, S., Adamson, J., Lees, A.J., Rossor, M.N., Quinn, N.P., Kertesz, A., et al., 2001. Corticobasal degeneration and progressive supranuclear palsy share a common tau haplotype. Neurology 56, 1702–1706.
Houlden, H., Schneider, S.A., Paudel, R., Melchers, A., Schwingenschuh, P., Edwards, M., Hardy, J., Bhatia, K.P., 2010. THAP1 mutations (DYT6) are an additional cause of early-onset dystonia. Neurology 74, 846–850.
Hutchins, J.B., Casagrande, V.A., 1989. Vimentin: changes in distribution during brain development. Glia 2, 55–66.
Ichinose, H., Ohye, T., Takahashi, E., Seki, N., Hori, T., Segawa, M., Nomura, Y., Endo, K., Tanaka, H., Tsuji, S., et al., 1994. Hereditary progressive dystonia with marked diurnal fluctuation caused by mutations in the GTP cyclohydrolase I gene. Nat. Genet. 8, 236–242.
Jinnah, H.A., Ceballos-Picot, I., Torres, R.J., Visser, J.E., Schretlen, D.J., Verdu, A., Larovere, L.E., Chen, C.J., Cossu, A., Wu, C.H., et al., 2010. Attenuated variants of Lesch-Nyhan disease. Brain 133, 671–689.
Kabakci, K., Hedrich, K., Leung, J.C., Mitterer, M., Vieregge, P., Lencer, R., Hagenah, J., Garrels, J., Witt, K., Klostermann, F., et al., 2004. Mutations in DYT1: extension of the phenotypic and mutational spectrum. Neurology 62, 395–400.

Kabakci, K., Isbruch, K., Schilling, K., Hedrich, K., de Carvalho Aguiar, P., Ozelius, L.J., Kramer, P.L., Schwarz, M.H., Klein, C., 2005. Genetic heterogeneity in rapid onset dystonia-parkinsonism: description of a new family. J. Neurol. Neurosurg. Psychiatry 76, 860–862.
Kamm, C., Boston, H., Hewett, J., Wilbur, J., Corey, D.P., Hanson, P.I., Ramesh, V., Breakefield, X.O., 2004. The early onset dystonia protein torsinA interacts with kinesin light chain 1. J. Biol. Chem. 279, 19882–19892.
Khan, N.L., Graham, E., Critchley, P., Schrag, A.E., Wood, N.W., Lees, A.J., Bhatia, K.P., Quinn, N., 2003a. Parkin disease: a phenotypic study of a large case series. Brain 126, 1279–1292.
Khan, N.L., Wood, N.W., Bhatia, K.P., 2003b. Autosomal recessive, DYT2-like primary torsion dystonia: a new family. Neurology 61, 1801–1803.
Kim, H.T., Edwards, M.J., Tyson, J., Quinn, N.P., Bitner-Glindzicz, M., Bhatia, K.P., 2007. Blepharospasm and limb dystonia caused by Mohr-Tranebjaerg syndrome with a novel splice-site mutation in the deafness/dystonia peptide gene. Mov. Disord. 22, 1328–1331.
Kindy, M.S., Bhat, A.N., Bhat, N.R., 1992. Transient ischemia stimulates glial fibrillary acid protein and vimentin gene expression in the gerbil neocortex, striatum and hippocampus. Mol. Brain Res. 13, 199–206.
Klein, C., Liu, L., Doheny, D., Kock, N., Muller, B., de Carvalho Aguiar, P., Leung, J., de Leon, D., Bressman, S.B., Silverman, J., et al., 2002. Epsilon-sarcoglycan mutations found in combination with other dystonia gene mutations. Ann. Neurol. 52, 675–679.
Konakova, M., Huynh, D.P., Yong, W., Pulst, S.M., 2001. Cellular distribution of torsin A and torsin B in normal human brain. Arch. Neurol. 58, 921–927.
Konakova, M., Pulst, S.M., 2001. Immunocytochemical characterization of torsin proteins in mouse brain. Brain Res. 922, 1–8.
Kuczmarski, R.J., Ogden, C.L., Grummer-Strawn, L.M., Flegal, K.M., Guo, S.S., Wei, R., Mei, Z., Curtin, L.R., Roche, A.F., Johnson, C.L., 2000. CDC growth charts: United States. Adv. Data 314, 1–27.
Kuner, R., Teismann, P., Trutzel, A., Naim, J., Richter, A., Schmidt, N., Bach, A., Ferger, B., Schneider, A., 2004. TorsinA, the gene linked to early-onset dystonia, is upregulated by the dopaminergic toxin MPTP in mice. Neurosci. Lett. 355, 126–130.
Kuner, R., Teismann, P., Trutzel, A., Naim, J., Richter, A., Schmidt, N., von Ahsen, O., Bach, A., Ferger, B., Schneider, A., 2003. TorsinA protects against oxidative stress in COS-1 and PC12 cells. Neurosci. Lett. 350, 153–156.
Larnaout, A., Belal, S., Miladi, N., Kaabachi, N., Mebazza, A., Dhondt, J.L., Hentati, F., 1998. Juvenile form of dihydropteridine reductase deficiency in 2 Tunisian patients. Neuropediatrics 29, 322–323.
Le Ber, I., Clot, F., Vercueil, L., Camuzat, A., Viemont, M., Benamar, N., De Liege, P., Ouvrard-Hernandez, A.M., Pollak, P., Stevanin, G., et al., 2006. Predominant dystonia with marked cerebellar atrophy: a rare phenotype in familial dystonia. Neurology 67, 1769–1773.
LeDoux, M.S., 2009. Meige syndrome: what's in a name? Parkinsonism Relat. Disord. 15, 483–489.
LeDoux, M.S., 2012a. Dystonia: phenomenology. Parkinsonism Relat Disord. 18(Supp 1), S162–164.
LeDoux, M.S., Xiao, J., Rudzińska, M., Bastian, R.W., Wszolek, Z.K., Van Gerpen, J.A., Puschmann, A., Momčilović, D., Vemula, S., Zhao, Y., 2012b. Genotype-phenotype correlations in THAP1 dystonia: molecular foundations and description of new cases. Parkinsonism Relat. Disord. 18, 414–425.
Lee, H.Y., Xu, Y., Huang, Y., Ahn, A.H., Auburger, G.W., Pandolfo, M., Kwiecinski, H., Grimes, D.A., Lang, A.E., Nielsen, J.E., et al., 2004. The gene for paroxysmal non-

kinesigenic dyskinesia encodes an enzyme in a stress response pathway. Hum. Mol. Genet. 13, 3161–3170.

Lee, W.L., Tay, A., Ong, H.T., Goh, L.M., Monaco, A.P., Szepetowski, P., 1998. Association of infantile convulsions with paroxysmal dyskinesias (ICCA syndrome): confirmation of linkage to human chromosome 16p12-q12 in a Chinese family. Hum. Genet. 103, 608–612.

Leube, B., Hendgen, T., Kessler, K.R., Knapp, M., Benecke, R., Auburger, G., 1997. Sporadic focal dystonia in northwest Germany: molecular basis on chromosome 18p. Ann. Neurol. 42, 111–114.

Leube, B., Kessler, K.R., Ferbert, A., Ebke, M., Schwendemann, G., Erbguth, F., Benecke, R., Auburger, G., 1999. Phenotypic variability of the DYT1 mutation in German dystonia patients. Acta Neurol. Scand. 99, 248–251.

Leube, B., Rudnicki, D., Ratzlaff, T., Kessler, K.R., Benecke, R., Auburger, G., 1996. Idiopathic torsion dystonia: assignment of a gene to chromosome 18p in a German family with adult onset, autosomal dominant inheritance and purely focal distribution. Hum. Mol. Genet. 5, 1673–1677.

Leung, J.C., Klein, C., Friedman, J., Vieregge, P., Jacobs, H., Doheny, D., Kamm, C., DeLeon, D., Pramstaller, P.P., Penney, J.B., et al., 2001. Novel mutation in the *TOR1A* (DYT1) gene in atypical early onset dystonia and polymorphisms in dystonia and early onset parkinsonism. Neurogenetics 3, 133–143.

Li, J., Zhu, X., Wang, X., Sun, W., Feng, B., Du, T., Sun, B., Niu, F., Wei, H., Wu, X., et al., 2012. Targeted genomic sequencing identifies *PRRT2* mutations as a cause of paroxysmal kinesigenic choreoathetosis. J. Med. Genet. 49, 76–78.

Liu, Q., Qi, Z., Wan, X.H., Li, J.Y., Shi, L., Lu, Q., Zhou, X.Q., Qiao, L., Wu, L.W., Liu, X.Q., et al., 2012. Mutations in PRRT2 result in paroxysmal dyskinesias with marked variability in clinical expression. J. Med. Genet. 49, 79–82.

Lohmann, E., Koroglu, C., Hanagasi, H.A., Dursun, B., Tasan, E., Tolun, A., 2012a. A homozygous frameshift mutation of sepiapterin reductase gene causing parkinsonism with onset in childhood. Parkinsonism Relat. Disord. 18, 191–193.

Lohmann, K., Uflacker, N., Erogullari, A., Lohnau, T., Winkler, S., Dendorfer, A., Schneider, S.A., Osmanovic, A., Svetel, M., Ferbert, A., et al., 2012b. Identification and functional analysis of novel *THAP1* mutations. Eur. J. Hum. Genet. 20, 171–175.

Lüdecke, B., Dworniczak, B., Bartholomé, K., 1995. A point mutation in the tyrosine hydroxylase gene associated with Segawa's syndrome. Hum. Genet. 95, 123–125.

Makino, S., Kaji, R., Ando, S., Tomizawa, M., Yasuno, K., Goto, S., Matsumoto, S., Tabuena, M.D., Maranon, E., Dantes, M., et al., 2007. Reduced neuron-specific expression of the TAF1 gene is associated with X-linked dystonia-parkinsonism. Am. J. Hum. Genet. 80, 393–406.

Marras, C., Van den Eeden, S.K., Fross, R.D., Benedict-Albers, K.S., Klingman, J., Leimpeter, A.D., Nelson, L.M., Risch, N., Karter, A.J., Bernstein, A.L., et al., 2007. Minimum incidence of primary cervical dystonia in a multiethnic health care population. Neurology 69, 676–680.

McLean, P.J., Kawamata, H., Shariff, S., Hewett, J., Sharma, N., Ueda, K., Breakefield, X.O., Hyman, B.T., 2002. TorsinA and heat shock proteins act as molecular chaperones: suppression of alpha-synuclein aggregation. J. Neurochem. 83, 846–854.

McNeill, A., Birchall, D., Hayflick, S.J., Gregory, A., Schenk, J.F., Zimmerman, E.A., Shang, H., Miyajima, H., Chinnery, P.F., 2008. T2* and FSE MRI distinguishes four subtypes of neurodegeneration with brain iron accumulation. Neurology 70, 1614–1619.

Moretti, P., Hedera, P., Wald, J., Fink, J., 2005. Autosomal recessive primary generalized dystonia in two siblings from a consanguineous family. Mov. Disord. 20, 245–247.

Naismith, T.V., Heuser, J.E., Breakefield, X.O., Hanson, P.I., 2004. TorsinA in the nuclear envelope. Proc. Natl. Acad. Sci. U. S. A. 101, 7612–7617.

Nery, F.C., Zeng, J., Niland, B.P., Hewett, J., Farley, J., Irimia, D., Li, Y., Wiche, G., Sonnenberg, A., Breakefield, X.O., 2008. TorsinA binds the KASH domain of nesprins and participates in linkage between nuclear envelope and cytoskeleton. J. Cell Sci. 121, 3476–3486.
Neuwald, A.F., Aravind, L., Spouge, J.L., Koonin, E.V., 1999. AAA+: a class of chaperone-like ATPases associated with the assembly, operation, and disassembly of protein complexes. Genome Res. 9, 27–43.
Nolte, D., Niemann, S., Müller, U., 2003. Specific sequence changes in multiple transcript system DYT3 are associated with X-linked dystonia parkinsonism. Proc. Natl. Acad. Sci. U. S. A. 100, 10347–10352.
Norgren, N., Mattson, E., Forsgren, L., Holmberg, M., 2011. A high-penetrance form of late-onset torsion dystonia maps to a novel locus (DYT21) on chromosome 2q14.3-q21.3. Neurogenetics 12, 137–143.
Ogura, T., Wilkinson, A.J., 2001. AAA+ superfamily ATPases: common structure–diverse function. Genes Cells 6, 575–597.
Oliveira, J.R., Spiteri, E., Sobrido, M.J., Hopfer, S., Klepper, J., Voit, T., Gilbert, J., Wszolek, Z.K., Calne, D.B., Stoessl, A.J., et al., 2004. Genetic heterogeneity in familial idiopathic basal ganglia calcification (Fahr disease). Neurology 63, 2165–2167.
Opal, P., Tintner, R., Jankovic, J., Leung, J., Breakefield, X.O., Friedman, J., Ozelius, L., 2002. Intrafamilial phenotypic variability of the DYT1 dystonia: from asymptomatic TOR1A gene carrier status to dystonic storm. Mov. Disord. 17, 339–345.
Opladen, T., Hoffmann, G., Horster, F., Hinz, A.B., Neidhardt, K., Klein, C., Wolf, N., 2011. Clinical and biochemical characterization of patients with early infantile onset of autosomal recessive GTP cyclohydrolase I deficiency without hyperphenylalaninemia. Mov. Disord. 26, 157–161.
O'Riordan, S., Cockburn, D., Barton, D., Lynch, T., Hutchinson, M., 2002. Primary torsion dystonia due to the *Tor1A* GAG deletion in an Irish family. Ir. J. Med. Sci. 171, 31–32.
Ozelius, L.J., Hewett, J.W., Page, C.E., Bressman, S.B., Kramer, P.L., Shalish, C., de Leon, D., Brin, M.F., Raymond, D., Corey, D.P., et al., 1997. The early-onset torsion dystonia gene (DYT1) encodes an ATP-binding protein. Nat. Genet. 17, 40–48.
Ozelius, L.J., Page, C.E., Klein, C., Hewett, J.W., Mineta, M., Leung, J., Shalish, C., Bressman, S.B., de Leon, D., Brin, M.F., et al., 1999. The *TOR1A* (DYT1) gene family and its role in early onset torsion dystonia. Genomics 62, 377–384.
Parker, N., 1985. Hereditary whispering dysphonia. J. Neurol. Neurosurg. Psychiatry 48, 218–224.
Perlman, S.L., 2011. Spinocerebellar degenerations. Handb. Clin. Neurol. 100, 113–140.
Peters, G.A., Seachrist, D.D., Keri, R.A., Sen, G.C., 2009. The double-stranded RNA-binding protein, PACT, is required for postnatal anterior pituitary proliferation. Proc. Natl. Acad. Sci. U. S. A. 106, 10696–10701.
Poorkaj, P., Muma, N.A., Zhukareva, V., Cochran, E.J., Shannon, K.M., Hurtig, H., Koller, W.C., Bird, T.D., Trojanowski, J.Q., Lee, V.M., et al., 2002. An R5L tau mutation in a subject with a progressive supranuclear palsy phenotype. Ann. Neurol. 52, 511–516.
Puschmann, A., Xiao, J., Bastian, R.W., Searcy, J.A., LeDoux, M.S., Wszolek, Z.K., 2011. An African-American family with dystonia. Parkinsonism Relat. Disord. 17, 547–550.
Rainier, S., Thomas, D., Tokarz, D., Ming, L., Bui, M., Plein, E., Zhao, X., Lemons, R., Albin, R., Delaney, C., et al., 2004. Myofibrillogenesis regulator 1 gene mutations cause paroxysmal dystonic choreoathetosis. Arch. Neurol. 61, 1025–1029.
Reich, S.G., Grill, S.E., 2009. Corticobasal degeneration. Curr. Treat. Options Neurol. 11, 179–185.
Risch, N., de Leon, D., Ozelius, L., Kramer, P., Almasy, L., Singer, B., Fahn, S., Breakefield, X., Bressman, S., 1995. Genetic analysis of idiopathic torsion dystonia in Ashkenazi Jews and their recent descent from a small founder population. Nat. Genet. 9, 152–159.

Risch, N.J., Bressman, S.B., Senthil, G., Ozelius, L.J., 2007. Intragenic Cis and Trans modification of genetic susceptibility in DYT1 torsion dystonia. Am. J. Hum. Genet. 80, 1188–1193.

Ritz, K., van Schaik, B.D., Jakobs, M.E., van Kampen, A.H., Aronica, E., Tijssen, M.A., Baas, F., 2011. *SGCE* isoform characterization and expression in human brain: implications for myoclonus-dystonia pathogenesis? Eur. J. Hum. Genet. 19, 438–444.

Rodacker, V., Toustrup-Jensen, M., Vilsen, B., 2006. Mutations Phe785Leu and Thr618Met in Na^+, K^+-ATPase, associated with familial rapid-onset dystonia parkinsonism, interfere with Na+ interaction by distinct mechanisms. J. Biol. Chem. 281, 18539–18548.

Roussigne, M., Cayrol, C., Clouaire, T., Amalric, F., Girard, J.P., 2003. THAP1 is a nuclear proapoptotic factor that links prostate-apoptosis-response-4 (Par-4) to PML nuclear bodies. Oncogene 22, 2432–2442.

Sako, W., Morigaki, R., Kaji, R., Tooyama, I., Okita, S., Kitazato, K., Nagahiro, S., Graybiel, A.M., Goto, S., 2011. Identification and localization of a neuron-specific isoform of TAF1 in rat brain: implications for neuropathology of DYT3 dystonia. Neuroscience 189, 100–107.

Sancho-Tello, M., Valles, S., Montoliu, C., Renau-Piqueras, J., Guerri, C., 1995. Developmental pattern of GFAP and vimentin gene expression in rat brain and in radial glial cultures. Glia 15, 157–166.

Santangelo, G., 1934. Contributo clinico alla conoscenza delle forme familiari della dysbasia lordotica progressiva (spasmo di torsione). Psychiat. Neuropat. 62, 52–77.

Schapira, A.H., Warner, T., Gash, M.T., Cleeter, M.W., Marinho, C.F., Cooper, J.M., 1997. Complex I function in familial and sporadic dystonia. Ann. Neurol. 41, 556–559.

Schneider, S.A., Bhatia, K.P., 2008. Dystonia in the Woodhouse Sakati syndrome: a new family and literature review. Mov. Disord. 23, 592–596.

Schneider, S.A., Bhatia, K.P., 2010a. Rare causes of dystonia parkinsonism. Curr. Neurol. Neurosci. Rep. 10, 431–439.

Schneider, S.A., Bhatia, K.P., 2010b. Secondary dystonia–clinical clues and syndromic associations. Eur. J. Neurol. 17 (Suppl. 1), 52–57.

Schneider, S.A., Paisan-Ruiz, C., Garcia-Gorostiaga, I., Quinn, N.P., Weber, Y.G., Lerche, H., Hardy, J., Bhatia, K.P., 2009. GLUT1 gene mutations cause sporadic paroxysmal exercise-induced dyskinesias. Mov. Disord. 24, 1684–1688.

Schneider, S.A., Ramirez, A., Shafiee, K., Kaiser, F.J., Erogullari, A., Bruggemann, N., Winkler, S., Bahman, I., Osmanovic, A., Shafa, M.A., et al., 2011. Homozygous *THAP1* mutations as cause of early-onset generalized dystonia. Mov. Disord. 26, 858–861.

Segawa, M., Hosaka, A., Miyagawa, F., Nomura, Y., Imai, H., 1976. Hereditary progressive dystonia with marked diurnal fluctuation. Adv. Neurol. 14, 215–233.

Seibler, P., Djarmati, A., Langpap, B., Hagenah, J., Schmidt, A., Bruggemann, N., Siebner, H., Jabusch, H.C., Altenmuller, E., Munchau, A., et al., 2008. A heterozygous frameshift mutation in PRKRA (DYT16) associated with generalised dystonia in a German patient. Lancet Neurol. 7, 380–381.

Sengel, C., Gavarini, S., Sharma, N., Ozelius, L.J., Bragg, D.C., 2011. Dimerization of the DYT6 dystonia protein, THAP1, requires residues within the coiled-coil domain. J. Neurochem. 118, 1087–1100.

Sethi, K.D., Jankovic, J., 2002. Dystonia in spinocerebellar ataxia type 6. Mov. Disord. 17, 150–153.

Shamim, E.A., Chu, J., Scheider, L.H., Savitt, J., Jinnah, H.A., Hallett, M., 2011. Extreme task specificity in writer's cramp. Mov. Disord. 26, 2107–2109.

Shan, Y., Cortopassi, G., 2011. HSC20 interacts with frataxin and is involved in iron-sulfur cluster biogenesis and iron homeostasis. Hum. Mol. Genet. [Epub ahead of print] PubMed PMID: 22171070.

Sharma, N., Armata, I.A., Multhaupt-Buell, T.J., Ozelius, L.J., Xin, W., Sims, K.B., 2011. Mutation in 5' upstream region of GCHI gene causes familial dopa-responsive dystonia. Mov. Disord. 26, 2140–2141.
Sharma, N., Hewett, J., Ozelius, L.J., Ramesh, V., McLean, P.J., Breakefield, X.O., Hyman, B.T., 2001. A close association of torsinA and alpha-synuclein in Lewy bodies: a fluorescence resonance energy transfer study. Am. J. Path. 159, 339–344.
Shashidharan, P., Good, P.F., Hsu, A., Perl, D.P., Brin, M.F., Olanow, C.W., 2000. TorsinA accumulation in Lewy bodies in sporadic Parkinson's disease. Brain Res. 877, 379–381.
Shashidharan, P., Paris, N., Sandu, D., Karthikeyan, L., McNaught, K.S., Walker, R.H., Olanow, C.W., 2004. Overexpression of torsinA in PC12 cells protects against toxicity. J. Neurochem. 88, 1019–1025.
Shen, Y., Lee, H.Y., Rawson, J., Ojha, S., Babbitt, P., Fu, Y.H., Ptácek, L.J., 2011. Mutations in PNKD causing paroxysmal dyskinesia alters protein cleavage and stability. Hum. Mol. Genet. 20, 2322–2332.
Spacey, S.D., Adams, P.J., Lam, P.C., Materek, L.A., Stoessl, A.J., Snutch, T.P., Hsiung, G.Y., 2006. Genetic heterogeneity in paroxysmal nonkinesigenic dyskinesia. Neurology 66, 1588–1590.
Spacey, S.D., Valente, E.M., Wali, G.M., Warner, T.T., Jarman, P.R., Schapira, A.H., Dixon, P.H., Davis, M.B., Bhatia, K.P., Wood, N.W., 2002. Genetic and clinical heterogeneity in paroxysmal kinesigenic dyskinesia: evidence for a third EKD gene. Mov. Disord. 17, 717–725.
Spina, S., Murrell, J.R., Yoshida, H., Ghetti, B., Bermingham, N., Sweeney, B., Dlouhy, S.R., Crowther, R.A., Goedert, M., Keohane, C., 2007. The novel Tau mutation G335S: clinical, neuropathological and molecular characterization. Acta Neuropathol. 113, 461–470.
Steinberger, D., Blau, N., Goriuonov, D., Bitsch, J., Zuker, M., Hummel, S., Müller, U., 2004. Heterozygous mutation in 5'-untranslated region of sepiapterin reductase gene (SPR) in a patient with dopa-responsive dystonia. Neurogenetics 5, 187–190.
Swerdlow, R.H., Wooten, G.F., 2001. A novel deafness/dystonia peptide gene mutation that causes dystonia in female carriers of Mohr-Tranebjaerg syndrome. Ann. Neurol. 50, 537–540.
Swoboda, K.J., 2006. Disorders of amine biosynthesis. Future Neurol. 1, 605–614.
Szepetowski, P., Rochette, J., Berquin, P., Piussan, C., Lathrop, G.M., Monaco, A.P., 1997. Familial infantile convulsions and paroxysmal choreoathetosis: a new neurological syndrome linked to the pericentromeric region of human chromosome 16. Am. J. Hum. Genet. 61, 889–898.
Tarsy, D., Sweadner, K.J., Song, P.C., 2010. Case records of the Massachusetts General Hospital. Case 17-2010-a 29-year-old woman with flexion of the left hand and foot and difficulty speaking. New Engl. J. Med. 362, 2213–2219.
Thiffault, I., Rioux, M.F., Tetreault, M., Jarry, J., Loiselle, L., Poirier, J., Gros-Louis, F., Mathieu, J., Vanasse, M., Rouleau, G.A., et al., 2006. A new autosomal recessive spastic ataxia associated with frequent white matter changes maps to 2q33-34. Brain 129, 2332–2340.
Thöny, B., Blau, N., 2006. Mutations in the BH4-metabolizing genes GTP cyclohydrolase I, 6-pyruvoyl-tetrahydropterin synthase, sepiapterin reductase, carbinolamine-4a-dehydratase, and dihydropteridine reductase. Hum. Mutat. 27, 870–878.
Tolosa, E., Compta, Y., 2006. Dystonia in Parkinson's disease. J. Neurol. 253 (Suppl. 7), VII7–VII13.
Tomita, H., Nagamitsu, S., Wakui, K., Fukushima, Y., Yamada, K., Sadamatsu, M., Masui, A., Konishi, T., Matsuishi, T., Aihara, M., et al., 1999. Paroxysmal kinesigenic choreoathetosis locus maps to chromosome 16p11.2-q12.1. Am. J. Hum. Genet. 65, 1688–1697.

Tońska, K., Kodron, A., Bartnik, E., 2010. Genotype-phenotype correlations in Leber hereditary optic neuropathy. Biochim. Biophys. Acta 1797, 1119–1123.
Torres, G.E., Sweeney, A.L., Beaulieu, J.M., Shashidharan, P., Caron, M.G., 2004. Effect of torsinA on membrane proteins reveals a loss of function and a dominant-negative phenotype of the dystonia-associated DeltaE-torsinA mutant. Proc. Natl. Acad. Sci. U. S. A. 101, 15650–15655.
Uitti, R.J., Maraganore, D.M., 1993. Adult onset familial cervical dystonia: report of a family including monozygotic twins. Mov. Disord. 8, 489–494.
Vale, R.D., 2000. AAA proteins. Lords of the ring. J. Cell Biol. 150, F13–19.
Valente, E.M., Bentivoglio, A.R., Cassetta, E., Dixon, P.H., Davis, M.B., Ferraris, A., Ialongo, T., Frontali, M., Wood, N.W., Albanese, A., 2001. DYT13, a novel primary torsion dystonia locus, maps to chromosome 1p36.13–36.32 in an Italian family with cranial-cervical or upper limb onset. Ann. Neurol. 49, 362–366.
Valente, E.M., Spacey, S.D., Wali, G.M., Bhatia, K.P., Dixon, P.H., Wood, N.W., Davis, M.B., 2000. A second paroxysmal kinesigenic choreoathetosis locus (EKD2) mapping on 16q13-q22.1 indicates a family of genes which give rise to paroxysmal disorders on human chromosome 16. Brain 123, 2040–2045.
Valente, E.M., Warner, T.T., Jarman, P.R., Mathen, D., Fletcher, N.A., Marsden, C.D., Bhatia, K.P., Wood, N.W., 1998. The role of DYT1 in primary torsion dystonia in Europe. Brain 121, 2335–2339.
van Gaalen, J., Giunti, P., van de Warrenburg, B.P., 2011. Movement disorders in spinocerebellar ataxias. Mov. Disord. 26, 792–800.
Van Gerpen, J.A., LeDoux, M.S., Wszolek, Z.K., 2010. Adult-onset leg dystonia due to a missense mutation in THAP1. Mov. Disord. 25, 1306–1307.
Vander Heyden, A.B., Naismith, T.V., Snapp, E.L., Hanson, P.I., 2011. Static retention of the lumenal monotopic membrane protein torsinA in the endoplasmic reticulum. EMBO J. 30, 3217–3231.
Vasudevan, A., Breakefield, X.O., Bhide, P.G., 2006. Developmental patterns of torsinA and torsinB expression. Brain Res. 1073–1074, 139–145.
Verbeek, M.M., Steenbergen-Spanjers, G.C., Willemsen, M.A., Hol, F.A., Smeitink, J., Seeger, J., Grattan-Smith, P., Ryan, M.M., Hoffmann, G.F., Donati, M.A., et al., 2007. Mutations in the cyclic adenosine monophosphate response element of the tyrosine hydroxylase gene. Ann. Neurol. 62, 422–426.
Verrotti, A., D'Egidio, C., Agostinelli, S., Gobbi, G., 2012. Glut1 deficiency: when to suspect and how to diagnose? Eur. J. Paediatr. Neurol. 16, 3–9.
Waite, A., Tinsley, C.L., Locke, M., Blake, D.J., 2009. The neurobiology of the dystrophin-associated glycoprotein complex. Ann. Med. 41, 344–359.
Walker, R.H., Good, P.F., Shashidharan, P., 2003. TorsinA immunoreactivity in inclusion bodies in trinucleotide repeat diseases. Mov. Disord. 18, 1041–1044.
Waln, O., LeDoux, M.S., 2011. Blepharospasm plus cervical dystonia with predominant anterocollis: a distinctive subphenotype of segmental craniocervical dystonia? Tremor Other Hyperkinet. Mov. (N.Y.) 1 pii: 33.
Wang, J.L., Cao, L., Li, X.H., Hu, Z.M., Li, J.D., Zhang, J.G., Liang, Y., San, A., Li, N., Chen, S.Q., et al., 2011. Identification of PRRT2 as the causative gene of paroxysmal kinesigenic dyskinesias. Brain 134, 3493–3501.
Weber, Y.G., Kamm, C., Suls, A., Kempfle, J., Kotschet, K., Schule, R., Wuttke, T.V., Maljevic, S., Liebrich, J., Gasser, T., et al., 2011. Paroxysmal choreoathetosis/spasticity (DYT9) is caused by a GLUT1 defect. Neurology 77, 959–964.
Weber, Y.G., Lerche, H., 2009. Genetics of paroxysmal dyskinesias. Curr. Neurol. Neurosci. Rep. 9, 206–211.
Weber, Y.G., Storch, A., Wuttke, T.V., Brockmann, K., Kempfle, J., Maljevic, S., Margari, L., Kamm, C., Schneider, S.A., Huber, S.M., et al., 2008. GLUT1 mutations

are a cause of paroxysmal exertion-induced dyskinesias and induce hemolytic anemia by a cation leak. J. Clin. Invest. 118, 2157–2168.

Weiss, E.M., Hershey, T., Karimi, M., Racette, B., Tabbal, S.D., Mink, J.W., Paniello, R.C., Perlmutter, J.S., 2006. Relative risk of spread of symptoms among the focal onset primary dystonias. Mov. Disord. 21, 1175–1181.

Weraarpachai, W., Antonicka, H., Sasarman, F., Seeger, J., Schrank, B., Kolesar, J.E., Lochmuller, H., Chevrette, M., Kaufman, B.A., Horvath, R., et al., 2009. Mutation in *TACO1*, encoding a translational activator of COX I, results in cytochrome c oxidase deficiency and late-onset Leigh syndrome. Nat. Genet. 41, 833–837.

Werner, E.R., Blau, N., Thöny, B., 2011. Tetrahydrobiopterin: biochemistry and pathophysiology. Biochem. J. 438, 397–414.

Wider, C., Melquist, S., Hauf, M., Solida, A., Cobb, S.A., Kachergus, J.M., Gass, J., Coon, K.D., Baker, M., Cannon, A., et al., 2008. Study of a Swiss dopa-responsive dystonia family with a deletion in GCH1: redefining DYT14 as DYT5. Neurology 70, 1377–1383.

Wilcox, R.A., Winkler, S., Lohmann, K., Klein, C., 2011. Whispering dysphonia in an Australian family (DYT4): a clinical and genetic reappraisal. Mov. Disord. 26, 2404–2408.

Willemsen, M.A., Verbeek, M.M., Kamsteeg, E.J., de Rijk-van Andel, J.F., Aeby, A., Blau, N., Burlina, A., Donati, M.A., Geurtz, B., Grattan-Smith, P.J., et al., 2010. Tyrosine hydroxylase deficiency: a treatable disorder of brain catecholamine biosynthesis. Brain 133, 1810–1822.

Wu, Y.R., Lee-Chen, G.J., Lang, A.E., Chen, C.M., Lin, H.Y., Chen, S.T., 2004. Dystonia as a presenting sign of spinocerebellar ataxia type 1. Mov. Disord. 19, 586–587.

Xiao, J., Bastian, R.W., Perlmutter, J.S., Racette, B.A., Tabbal, S.D., Karimi, M., Paniello, R.C., Blitzer, A., Batish, S.D., Wszolek, Z.K., et al., 2009. High-throughput mutational analysis of *TOR1A* in primary dystonia. BMC Med. Genet. 10, 24.

Xiao, J., Gong, S., Zhao, Y., LeDoux, M.S., 2004. Developmental expression of rat torsinA transcript and protein. Brain Res. Dev. Brain Res. 152, 47–60.

Xiao, J., LeDoux, M.S., 2003. Cloning, developmental regulation and neural localization of rat epsilon-sarcoglycan. Mol. Brain Res. 119, 132–143.

Xiao, J., Uitti, R.J., Zhao, Y., Vemula, S.R., Perlmutter, J.S., Wszolek, Z.K., Maraganore, D.M., Auburger, G., Leube, B., Lehnhoff, K., et al., 2012. Mutations in CIZ1 cause adult-onset primary cervical dystonia. Ann. Neurol. 71, 458–469.

Xiao, J., Zhao, Y., Bastian, R.W., Perlmutter, J.S., Racette, B.A., Tabbal, S.D., Karimi, M., Paniello, R.C., Wszolek, Z.K., Uitti, R.J., et al., 2010. Novel THAP1 sequence variants in primary dystonia. Neurology 74, 229–238.

Xiao, J., Zhao, Y., Bastian, R.W., Perlmutter, J.S., Racette, B.A., Tabbal, S.D., Karimi, M., Paniello, R.C., Wszolek, Z.K., Uitti, R.J., et al., 2011. The c.-237_236GA>TT THAP1 sequence variant does not increase risk for primary dystonia. Mov. Disord. 26, 549–552.

Zhao, Y., Xiao, J., Ueda, M., Wang, Y., Hines, M., Nowak Jr., T.S., LeDoux, M.S., 2008. Glial elements contribute to stress-induced torsinA expression in the CNS and peripheral nervous system. Neuroscience 155, 439–453.

Zimprich, A., Grabowski, M., Asmus, F., Naumann, M., Berg, D., Bertram, M., Scheidtmann, K., Kern, P., Winkelmann, J., Muller-Myhsok, B., et al., 2001. Mutations in the gene encoding epsilon-sarcoglycan cause myoclonus-dystonia syndrome. Nat. Genet. 29, 66–69.

Zirn, B., Grundmann, K., Huppke, P., Puthenparampil, J., Wolburg, H., Riess, O., Müller, U., 2008a. Novel TOR1A mutation p.Arg288Gln in early-onset dystonia (DYT1). J. Neurol. Neurosurg. Psychiatry 79, 1327–1330.

Zirn, B., Steinberger, D., Troidl, C., Brockmann, K., von der Hagen, M., Feiner, C., Henke, L., Müller, U., 2008b. Frequency of GCH1 deletions in dopa-responsive dystonia. J. Neurol. Neurosurg. Psychiatry 79, 183–186.

CHAPTER THREE

Exome Sequencing and Advances in Crop Improvement

Devi Singh[*,1] **Pankaj K. Singh**[†,1] **Sarika Chaudhary**[‡,1] **Kamiya Mehla**[†,1] **and Shashi Kumar**[¶,1]

[*]Molecular Biology Laboratory, Department of Genetics and Plant Breeding, Sardar Vallabhbhai Patel University of Agriculture and Technology, Meerut, UP, India
[†]Eppley Institute for Research in Cancer and Allied Diseases, University of Nebraska Medical Center, Omaha, NE, USA
[‡]UCSF Helen Diller Family Comprehensive Cancer Center, San Francisco, CA, USA
[¶]International Center for Genetic Engineering and Biotechnology, Aruna Asaf Ali Marg, New Delhi, India 110067

Contents

[1] These authors contributed equally to this work.

Advances in Genetics, Volume 79
ISSN 0065-2660,
http://dx.doi.org/10.1016/B978-0-12-394395-8.00003-7

Abstract

Next-generation sequencing strategies have opened new vistas for molecular plant breeding. The sequence information obtained by the advent of next-generation sequencing provides a valuable tool not only for improving domesticated crops but also for investigating the natural evolution of crops. Such information provides an enormous potential for sustainable agriculture. In this review, we discuss how such sequencing approaches have transformed exome sequencing into a practical utility that has enormous potential for crop improvement in agriculture. Furthermore, we also describe the future of crop improvement beyond the exome sequencing strategies.

I. INTRODUCTION

Traditional breeding strategies have been exhausted for crop improvement. Still, to meet the energy needs of the burgeoning world population, we need to further improve food production. Molecular marker-assisted breeding strategies have further facilitated crop quality improvement. However, even molecular marker-assisted breeding strategies have their own limitations, mostly dealing with the identification of proper biomarkers for various phenotypic traits. With the advancements in next-generation sequencing strategies, whole genomes can be sequenced with a relative ease, thus, facilitating the identification of new molecular markers or even replacing polymerase chain reaction (PCR) or restriction sequence-based biomarker screening with exome sequencing.

Although whole-genome sequencing approaches can identify sequence variations to a single nucleotide sequence, the cost and time for the whole sequence identification and further whole-genome assembly and analysis can be enormous. Furthermore, due to lack of any reference sequences, determining the accuracy of such identified sequences can be a challenging task.

Hence, from a practical standpoint, exome sequencing can alleviate such issues. Exome refers to the entire protein-coding sequence in an organism. Although exome sequences only represent almost a hundredth of the whole genome and do not provide a complete picture of gene regulation, they represent all the sequences that code for proteins, which regulate phenotypes. A comparison of the whole-genome and exome-based sequencing strategies is illustrated in Table 3.1. Exome sequencing has proven to be an elegant genetic tool for dissecting the molecular basis of diseases and traits.

II. SEQUENCING: TOOLS AND TECHNIQUES

A. A Brief Historical Perspective

Sequencing the genetic code is the heart of molecular biology and is must for identifying the molecular basis of various traits and diseases. Early attempts at sequencing the DNA were made by Allan Maxam and Walter Gilbert who developed a chemical modification–dependent cleavage method (also known as chemical sequencing method) for DNA sequencing in 1976–1977 (Maxam

Table 3.1 Comparison of whole-genome and exome-sequencing methods

Feature	Whole-genome sequencing	Exome sequencing
Sequence included	Whole genome	Only the sequences that code for proteins
Sequence size	Whole genome	Smaller than the whole genome due to smaller coding region
Time	Fairly long	Relatively faster due to smaller size
Cost	Expensive	Relatively cheaper
Assembly success rate	Low due to highly repetitive sequences	Most exomes can be assembled by sequence comparisons with other species
Ease of analysis	Low due to large sequence data size	High due to smaller sequence data size
Information excluded	No sequence information is excluded	All the noncoding regions excluded hence, the effects of miRNA on expression cannot be determined

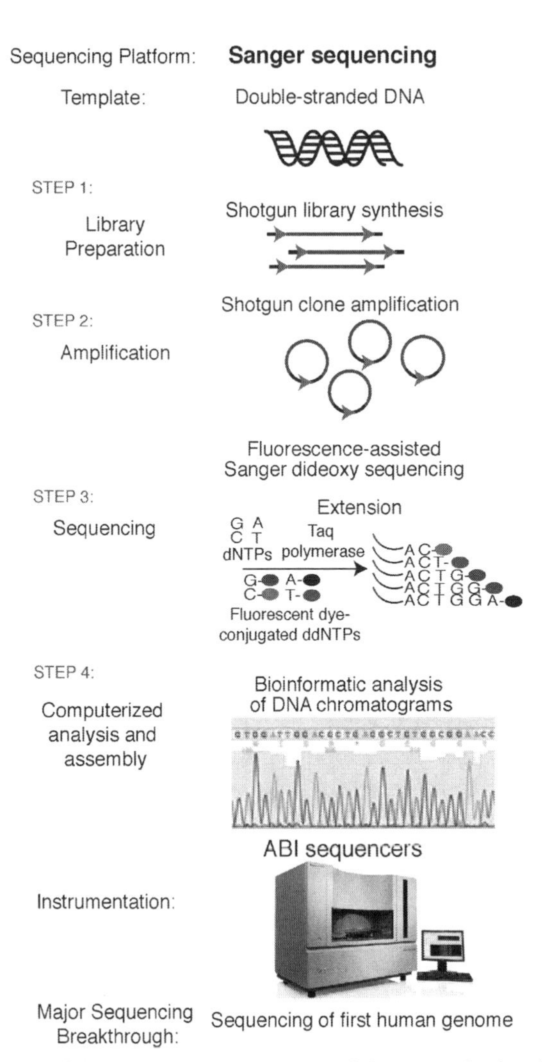

Figure 3.1 ***Sanger's dideoxy sequencing.*** Sanger's dideoxy method is the basis of the first generation high-throughput sequencers from Applied Biosystems. The template sequence is determined by fluorescent signals from incorporated fluorescent dideoxy nucleotides. For color version of this figure, the reader is referred to the online version of this book.

and Gilbert, 1977). Another early method that became very popular and is still in practice is Sanger's dideoxy chain termination sequencing method (Sanger and Coulson, 1975). This method requires the incorporation of a dideoxy nucleotide in a growing nucleotide chain that blocks chain elongation because of a missing 3'-OH group. The bands can be visualized on an SDS-PAGE by utilizing radioactive phosphorus containing nucleotides. Sanger's sequencing method was more efficient and it required fewer toxic chemicals. Later on, sequencing was made automated by performing Sanger's sequencing method with primers that were labeled with a fluorescent dye at the 5' terminus. Automation was a huge breakthrough in the DNA sequencing field because it provided ease, reliability, and cost-effectiveness in performing DNA sequencing. Later advancements in Sanger's dideoxy sequencing method included utilizing all four ddNTPs labeled with different fluorescent dyes in a single reaction (Smith *et al.*, 1986). Resolving the amplicons by capillary electrophoresis followed by LASER-based detection of the incorporated ddNTP in amplicons was another significant advancement. This method is much faster in contrast to a four-reaction sequencing method and hence is the method of choice for presently automated sequencing instrumentations (Fig. 3.1). Despite all the precision, this method has several limitations for high-throughput sequencing, including poor quality at the initial 20–50 bases, inability to clearly read sequences after 600–1000 bases due to poor size resolution of large-sized DNA fragments by capillary electrophoresis, nonspecific primer binding, and secondary structures in DNA. Initial attempt at sequencing larger fragments involved cloning smaller fragments in plasmids prior to sequencing. However, such methods also led to contamination by vector sequences. Improvements in bioinformatics along with sequencing strategies incorporating PCR-cloned fragments alleviated vector contamination issues. Recent methods involving combined amplification further improved longer reads (Murphy *et al.*, 2005; SenGupta and Cookson, 2010). The first automated fluorescent DNA sequencer utilized a paired-end sequencing approach to sequence the entire hypoxanthineguanine phosphoribosyltransferase (HPRT) gene (Edwards *et al.*, 1990; Pareek *et al.*, 2011). A slab gel electrophoresis system was utilized by ABI Prism 310, the first commercial DNA sequencer, in 1996. Later, the tiring slab gel pouring system was replaced by automated capillary reloading in the 96 capillary ABI Prism 3700. ABI Prism 3700 was utilized in sequencing the first human genome in 2003.

B. Next-Generation Sequencing Technology

To meet the ever-growing large sequence demands, several new automated sequencing methods have been developed and commercialized in recent years. Such ultrahigh-throughput sequencing platforms that do not utilize Sanger's dideoxy chain termination sequencing technology are collectively categorized as next-generation or second generation sequencing platforms. Genome Sequencer from Roche/454, Genome Analyzer from Illumina/Solexa, SOLiD™ from Applied Biosystems, and

Polonator from Dover Systems are some of the commercially available next-generation sequencers.

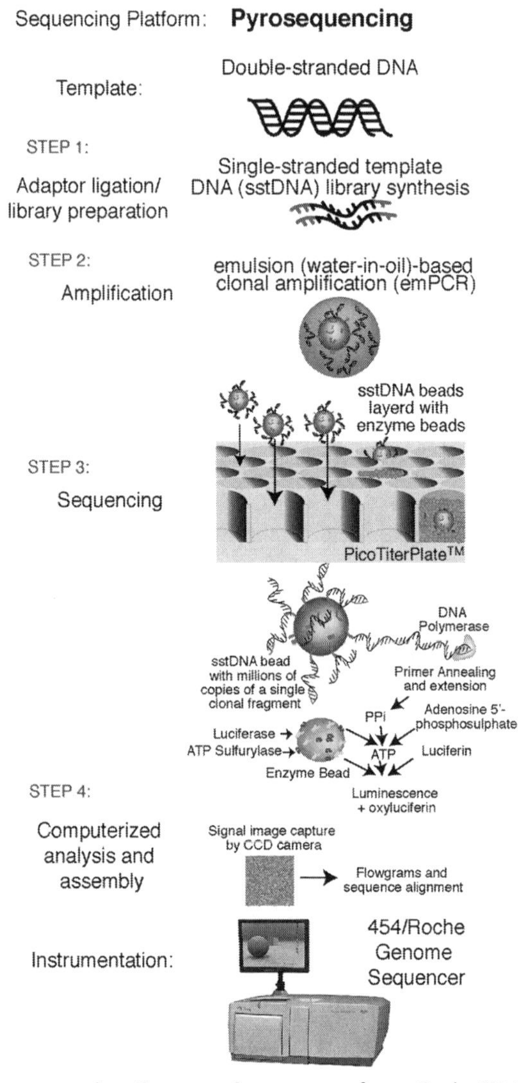

Figure 3.2 ***Pyrosequencing.*** Genome Sequencers from Roche/454 utilize pyrosequencing, where luminescent signals mark incorporation of bases into the growing single-stranded template on beads. For color version of this figure, the reader is referred to the online version of this book.

1. Pyrosequencing

Genome Sequencers from Roche/454 utilize the pyrosequencing platform, which is based on sequencing-by-synthesis (SBS) approach (http://www.lifesequencing.com) (Fig. 3.2). This platform first requires preparation of the DNA library with fragmented genomic DNA (gDNA) (300–500 bp fragments) (Margulies *et al.*, 2005). The fragments are then blunted and ligated at both ends with short adaptors, which serve as primers for the amplification of the fragments. 5'-biotin tag on one of the adapters enables amplicon immobilization on streptavidin-conjugated beads. Nick repair releases the nonbiotinylated strand to generate the single-stranded template DNA (sstDNA) library. Multiple samples (up to 12) can also be pooled together by utilizing unique barcoded adaptors. After titrating for optimal quantity and quality, the sstDNA library is then immobilized onto beads utilized for emPCR™ (emulsion-based PCR). The library-immobilized beads are then emulsified along with PCR reagents in water-in-oil emulsions. Each bead carrying a single amplicon is then clonally amplified to obtain millions of copies of the same single amplicon. sstDNA library beads are incubated with polymerase and enzyme beads, which contain immobilized ATP sulfurylase and luciferase enzymes, in a PicoTiterPlate™ device. This process ensures that each well contains only one sstDNA library bead. PicoTiterPlate™ is then placed into the instrument for pyrosequencing where the plate is sequentially layered with sequencing reagents and individual nucleotides by a fluidics system, ensuring parallel sequencing for millions of copies of sstDNA. If a complementary nucleotide is encountered in a nucleotide run, polymerase will extend the growing polynucleotide chain and hence cause the release of an inorganic pyrophosphate (PPi). Along with Adenosine 5'-phosphosulphate (APS), PPi serves as a substrate for ATP sulfurylase on the enzyme bead and gets converted into ATP. ATP thus produced is then utilized by luciferase enzyme on the enzyme bead that converts luciferin into oxyluciferin and generates luminescence. The luminescent signal is then detected by a CCD camera in the sequencer, where the signal strength indicates the number of nucleotides incorporated in a single flow. Bioinformatic analysis is performed for sequence assembly. The developments in the reaction chemistry of GS FLX systems now allow read lengths of up to 1 kb. GS20 was the first next-generation sequencer to be commercialized in 2005. Genome sequence of James Watson was determined in 2007 (project Jim) by utilizing this sequencing platform (Wheeler *et al.*, 2008). Also, this platform was successfully utilized to sequence the Neanderthal genome (Green *et al.*, 2010).

2. Reversible terminator-based sequencing

This platform is utilized by the Illumina/Solexa Genome Analyzers (http://www.illumina.com). It utilizes an optically transparent surface to which randomly fragmented gDNA molecules are attached. Oligo-primed DNA fragments are extended by simultaneous incubation with all four nucleotides

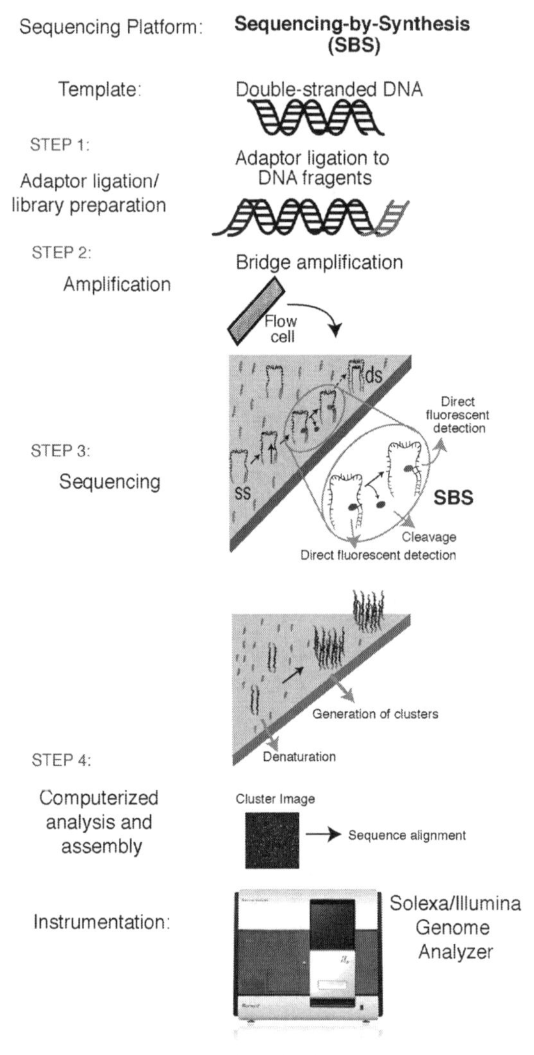

Figure 3.3 ***Sequencing-by-synthesis.*** Illumina/Solexa genome analyzers utilize sequencing-by-synthesis technology, where newly incorporated bases are detected by fluorescence imaging. For color version of this figure, the reader is referred to the online version of this book.

and polymerase in flow-cell channels. Bridge amplification causes strand extension for sequencing and generating a flow cell with ~108 clusters. Each cluster contains the same template with around a thousand copies. The clustered templates are sequenced by SBS technology, utilizing removable fluorescent dye-conjugated reversible terminators (Fig. 3.3). Sequence is determined by imaging a fluorescently labeled terminator during the addition of each dNTP and the terminator is then cleaved off to allow further base incorporation. This technology generates true sequences with high accuracy while minimizing context-specific errors. Laser excitation, in combination with total internal reflection (TIR) optics, is utilized to obtain fluorescence detection with high sensitivity. This technology allows sequential sequencing from both the strands. Once the first strand is sequenced, the template is regenerated to allow a second round of sequencing (75+ base read) from the opposite end by utilizing the paired-end module, which allows template regeneration (the complementary strand of the original template) and amplification.

3. *Sequencing-by-ligation*

Sequencing by Oligonucleotide Ligation and Detection (SOLiD™) was developed by ABI/Life Technologies (http://www.appliedbiosystems.com). This technology allows up to billions of short sequence reads at a time (Mitsui *et al.*, 2010). As a first step, fragment or mate-paired libraries are generated that are utilized by SOLiD™ systems. Clonal bead populations are generated by emulsion PCR in microreactors and the templates are then denatured to enrich beads with template extensions. The templates on enriched beads are subjected to a 3' modification to ensure attachment to the slide by covalent bonding. The 3' modified beads are deposited onto a glass slide in up to eight sections to accommodate for high bead densities on each slide. This allows a higher throughput in the same system. After primer–adapter hybridization on the template, ligation proceeds on the sequencing primer through competition between four distinctly fluorescent di-base probes. The ligated di-base is fluorescently detected and the remainder of the probe is then cleaved off. The template is subjected to a multiple cycles of such ligation, detection, and cleavage cycles. Each sequence tag undergoes five rounds of primer resets. By utilizing different primer sets, each sequencing position is interrogated by two different primers in two independent ligation reactions. The exact call chemistry with dual interrogation allows for up to 99.99% accuracy with SOLiD™ sequencing. The maximum read length for SOLiD™ systems is 2 × 60 bp for mate-paired libraries and 75 × 35 bp for paired-end libraries. These systems can utilize up to 1–12 independent lanes in two FlowChips, providing 10–15 Gb/day throughputs. Unlike the Roche/454 systems, the sequencing by ligation chemistry of SOLiD™ systems allows sequencing in the difficult homopolymer repeat regions. Polonator™ from Dover systems is another sequencer that utilizes sequencing-by-ligation platform.

4. Ion semiconductor-based nonoptical sequencing

This sequencing platform detects the hydrogen ions released during DNA polymerization. This is also an SBS approach. Ion PGM™ sequencer, a small-scale sequencer, and Ion Proton Sequencer, a large-scale sequencer capable of sequencing genomes, marketed by Ion Torrent Inc/Life Technologies, are essentially bench top next-generation sequencers that utilize this technology. These sequencers utilize micromachined wells, each holding a different DNA template, arranged in a high-density array (Rothberg *et al.*, 2011). An ion-sensitive layer is sandwiched between the well array and the ion sensor array. Each time a nucleotide is incorporated into a DNA strand, a hydrogen ion is released. The charge from the ion changes the pH of the solution in the well that is detected by the ion sensor. The chip is flooded sequentially with different nucleotides and no voltage change is recorded for a mismatch. A double voltage change is recorded for two identical bases. The major significance of this technology is fast speed and low upfront/operating costs. Each run goes for about an hour, utilizing 4 seconds per measurements and sequencing 100–200 nucleotides per run. However, very long homopolymers may be hard to sequence with this system.

5. DNA nanoball sequencing

This sequencing platform is utilized by Complete Genomics (www.completegenomics.com). It utilizes rolling circle replication, where small gDNA fragments are amplified in a manner so that all copies are unidirectionally connected together (Drmanac *et al.*, 2010; Porreca, 2010). Each such long single molecule is packaged into DNA nanoballs (DNBs) of ~200 nm diameters. Each DNB contains a 70-base long DNA fragment in hundreds of copies. DNA is then adhered to silicon chip at the desired spots. Each DNB is then sequenced by utilizing a combinatorial probe anchor-ligation (cPAL) technology, where a ligase enzyme attaches fluorescent molecules to nucleotides in DNBs. The sequence of nucleotides in DNBs is determined by fluorescence imaging utilizing a high-resolution CCD camera, reading the 70-base fragment, 5-base at a time. This technology allows defined (nonrandom, unlike other sequencing technologies) attachment of one DNB to each pit, thus allowing maximum reads from each flow cell. Furthermore, this technology does not require probe ligation to complete before the next round, thus minimizing reading errors. This technology utilizes a high fidelity DNA polymerase, minimizing inaccurate rolling circle amplification. Also, the hundreds of copies on a small spot intensify fluorescent signal and the fluorophore, being at a distant ligation point, do not affect ligation. However, short read lengths can give erroneous sequencing results, especially if such reads are present at multiple sequencing locations in the same genome. Also, multiple PCR rounds may introduce PCR bias.

C. Next-Next Generation Sequencing Technology

Further advancements in sequencing technologies have provided single molecule sequencing, also known as next-next or third generation sequencing technology (TGS). Some of the TGS sequencing platforms are discussed below.

1. True single molecule sequencing (TSMS)

This sequencing platform is utilized by Helicos Bioscience Corporation in the Heliscope™ sequencers. DNA sample is fragmented into short fragments of 100–200 bp (Thompson and Steinmann, 2010). A polyA tail adapter is added to the 3' end of each fragment followed by addition of a fluorescent adenosine nucleotide. These fragments are hybridized to a Helicos flow cell that contains billions of oligo-T universal capture sites immobilized to the flow cell. Each individual template is then sequenced at a very high density (100 million/cm^2) or billions per run. Posthybridization flow cells are loaded onto the instrument and the fluorescent templates are illuminated by laser. CCD camera produces multiple images in stepwise patterns to determine the location of each of the templates. Once imaged, the label is then cleaved and washed away. The sequencing reaction is then begun in sequential reactions by DNA polymerase in the presence of fluorescently labeled dNTPs (only one of dATP, dGTP, dCTP, or dTTP at a time). After each single nucleotide incorporation event, the excess DNA polymerase and dNTPs are then washed away and the camera then images the fluorescent label incorporation. The fluorescent label is then cleaved off and the other dNTPs are utilized in subsequent steps until desired read lengths are obtained. Images from each incorporation step are then analyzed to obtain the complete sequence. A significant advantage of this platform is the lack of any amplification steps. Hence, every strand is unique and its sequencing reaction is independent of the sequencing reactions on the other strands. With recent improvements, up to a billion bases can be sequenced per hour by utilizing this platform.

2. Single molecule, real-time (SMRT) sequencing

SMRT sequencing platform is utilized in PacBio *RS* sequencers by Pacific Biosciences (http://www.pacificbiosciences.com/products/smrt-technology). This technology requires generation of SMRTbell library (Korlach *et al.*, 2010). This is essentially an SBS technology where nucleotide incorporation is detected in real-time (Eid *et al.*, 2009). The library is created by generation of DNA fragments followed by end repair and hairpin adapter ligations to obtain circular SMRT DNA templates. The fragment size may vary from 250 bp to 10 kb. The library is then bound to DNA polymerase and DNA sequencing is performed on SMRT cells, which contain an array of close to 75,000 zero-mode waveguides (ZMWs). ZMWs are tens-of-nanometers-sized holes in a 100 nm metal film layered on a glass substrate. Each ZMW contains a single DNA polymerase on the glass surface. Whenever a fluorescently labeled nucleotide enters the bottom 30 nm of the ZMW, a fluorescence pulse is detected. If the nucleotide diffuses out of the ZMW, the pulse is short. But if the nucleotide is incorporated into the growing chain, the pulse will be for a longer time period. The fluorescence color determines which nucleotide is incorporated, while a longer pulse width, as compared to free diffusion, indicates nucleotide incorporation into the DNA. The DNA polymerase cleaves the nucleotide's terminal phosphate linked fluorophore (not the usual base linked) before translocating to the next base on the template. This technology can also detect DNA methylation, which is measured by five times longer interpulse duration in contrast to that of the unmethylated base (Flusberg *et al.*, 2010). This technology allows longer read lengths, usually greater than 3000 bp, that allow easier mapping and assembly. The sequencing reactions are very fast and the usual instrument time is close to 30 min.

3. Single-molecule RNAP motion-based real-time sequencing

This sequencing technique utilizes RNA polymerase (RNAP)-mediated transcription (Greenleaf and Block, 2006). A polystyrene bead is attached each to the RNAP enzyme and the distal end of DNA template. RNAP transcriptional motion along the template reduces the distance between the two beads. Both beads are placed in optical traps. Such displacements are measured with high precision at single-base pair resolution. One of the four dNTPs is present in a limiting concentration that makes the RNAP pause at every position that requires the limiting dNTP. Thus, four independent reactions, representing the limiting concentrations of each one of the four dNTPs, generate transcriptional position versus time information that is then aligned to obtain the final sequence. The read lengths for this platform are limited to the processivity of RNAP that is thousands of base pairs.

4. Nanopore sequencing

This sequencing platform is utilized by recently developed GridION and miniaturized MinION sequencers from Oxford Nanopore Technologies (www.nanoporetech.com). Nanopore sequencing technology, also known as "Strand sequencing," identifies individual nucleotide sequences as the DNA strand is passed through a membrane-inserted protein nanopore, one base at a time, by alterations in the ion current. The size, shape, and length of the DNA molecule determine the change in current. Each of the four passing nucleotides causes a distinctive variation in the current flow that allows identification of the nucleotide. By creating hairpins at the end of DNA fragment, this technology can sequence both sense and antisense of a DNA fragment. Hundreds of thousands of nanopores are built on an array chip that is used in these instruments. This platform can provide real-time sequencing of single DNA molecules at low cost and very fast pace, without damaging the DNA. The company is expected to start marketing 8000 nanopore-containing nodes by 2013 that will be have the capability to sequence the entire genome of a human being in 15 minutes. However, the technology currently has a 4% error rate that needs to be significantly reduced.

D. Technical Challenges to High-Throughput Sequencing

With the decreasing cost of high-throughput DNA sequencing technology and the availability of bench top sequencers, the technology has become affordable to the larger research community. Despite all the developments in the exome sequencing technology, utility of this technology for crop improvement still suffers from several roadblocks. There is an overgrowing need to create well-supported, thoroughly documented and robust computer algorithms/software solutions that can be utilized by researchers for large-scale genome sequence data analyses. Analysis of the massive amounts of data generated by the technology still remains a daunting task for most researchers. Even when several data analysis tools have been developed, their user-friendliness is still questionable. The alignment tools still need to be improved further to handle massive amounts of short reads. The next-generation/next-next generation sequencing technologies also need further improvements to increase accuracy and the ability to handle biases and genome polymorphisms. Despite its limitations, significant milestones have been reached by the next-generation/next-next generation sequencing technologies. Below we describe scopes and limitations of the exome sequencing technology in crop improvement.

III. EXOME SEQUENCING FOR CROP IMPROVEMENT

A. Exome Sequencing for Exploring Biodiversity

Traditionally, human beings have learned about the most morphological traits of plant and animal species from the wild and whatever suited our needs has been domesticated and further improved upon. Hence, biodiversity is essentially the natural source of all desirable traits we want in crops. Understanding how much diversity is present on the Earth, including the bottom of the ocean and how it is distributed is a great challenge. Furthermore, morphological analysis is not always adequate in identifying various closely resembling species, whereas genetic methods have frequently demonstrated the occurrence of species complexes or cryptic species. To expedite species identification, molecular taxonomists have assembled the Consortium for the Barcode of Life (CBOL) (www.barcoding.si.edu). To identify differences between species that can be amplified and sequenced to produce a species specific "barcode." Common interest is to identify a single gene that can be utilized to classify organisms into different species. Although the *mitochondrial gene cytochrome oxidase I (COI)* (Hebert *et al.*, 2003) is valuable in identifying some organisms, nuclear genes, especially those of the ribosomal RNA-encoding multigene family (18S & 28S rRNA), are useful in other cases. Also, the utility of *COI* in classifying plant genomes is limited, where chloroplast genome sequences are being evaluated for classification purposes.

Due to practical limitations of time and money, so far molecular identification of a species is mostly based on DNA fingerprinting. DNA fingerprinting relies upon the sequence modifications that can differentiate between species. This is achieved by designing species-specific primers that will only amplify DNA from a single or very closely related species. These primers can be linked to fluorescent tags. Alternatively, primers can be designed to amplify different sized products from several species of interest. The amplification products (amplicons) are then differentiated by molecular weight-dependent mobility upon electrophoresis. Another approach identifies different species by utilizing restriction enzymes to digest the PCR products from specimens. If the restriction sites and PCR primers are picked carefully, the digestion products can produce a unique "fingerprint" for several species. Other methods such as RAPD (Random Amplified Polymorphic DNA) fingerprinting can also be utilized for species-specific molecular fingerprinting for species identification. RAPD comprises a PCR reaction using two short random primers. DNA is amplified where the primers anneal to the genomic template sufficiently close to each other. The amplicons are then subjected to electrophoresis for visualization.

Sequence analysis can revolutionize biodiversity explorations and identification of new species and can thus prevent any miscategorizations. With the aid of next-generation sequencing techniques, it is now possible to perform exome sequencings on routine basis for revalidating/reconstructing the taxonomic classifications. Once all the exome data sets are available, exome sequence from any candidate species can be easily tested for homology with any other member of the same or related taxonomic classes. Since a single gene cannot serve as a master gene for barcoding, gene sequences can be compared across species within taxonomic classes to constantly monitor the optimal gene(s) for barcoding. Once the classification has been established, gene–trait relationships can be easily analyzed and any trait variations can be mapped to single base variations in the exonic regions by simply performing sequence comparisons.

B. Exome Sequencing to Study Host–Pathogen Interactions

Host–pathogen interactions result from constantly evolving dynamics of genomic interphases between plants and the corresponding pathogens. Virulence and susceptibility in the host–pathogen interactions can be altered by even single amino acid alterations in the regulatory proteins (Carroll *et al.*, 2011). Genetic engineering technologies have been successfully utilized for several aspects of pest control; for instance, Calvitti *et al.* (2010) established an *Aedes albopictus–Wolbachia* symbiotic association by artificial transfer of the wPip strain from *Culex pipiens* for effective pest control. Although some of the pathogen genomes are relatively small and can be sequenced with much ease, mapping the entire genomes of the host species for every occurrence of modification in host–pathogen interaction can be a challenging task. However, sequencing the exomes can serve as a cost-effective unbiased alternative and can provide valuable information on drafting strategies to combat disease pathogenesis. Deep transcriptome sequencing by utilizing Illumina's SBS technology has been successfully implemented to identify the genes involved in plant–fungal interactions (O'Brien *et al.*, 2010; Venu *et al.*, 2011). This was achieved by performing sequence analysis using BLAST against sequenced fungal genomes and rice genomic sequence to identify the genes involved in host–pathogen interactions that are expressed in both *S. homoeocarpa* mycelia and creeping bentgrass. The utility of next-generation sequencing technology has also been demonstrated in the evolution of plant pathogenesis in *Pseudomonas syringae*, where 25 new strains have been sequenced utilizing the next-generation sequencing technology (Lindeberg *et al.*, 2008). These studies indicated that *Pseudomonas syringae* genomes are highly dynamic, and extensive polymorphism was found in the distribution of type III secreted effectors (T3SEs) and other virulence-associated genes, even among strains within the same pathovar (Hu *et al.*, 2010). The genome sequences obtained by the next-generation methodology are still not finished or closed due, in part, to the difficulty of assembling the short reads produced by these methods through repetitive sequences, and in part to the lack of PCR-based genome closing efforts, which do not scale with the increased rate of shotgun sequencing (Hu *et al.*, 2010). Nonetheless, the information provided is valuable in understanding the virulence. Furthermore, the challenges in assembling the whole genomes will not even be an issue with the exome sequencing, which would in essence be sufficient for obtaining all such information.

C. Exome Sequencing for Investigating the Natural Evolution of Crops

Natural evolution of a number of crops has been investigated in detail by utilizing molecular markers. Studying the natural evolution of domesticated crops provides us with valuable information about the molecular events that can be mimicked artificially to facilitate the incorporation of desired traits into the crop plants. Evolutionary genomic approaches for identifying genes or genomic regions that have undergone selection have made possible the efficient and unbiased identification of genes involved in crop evolution

(Burke *et al.*, 2007). Such studies provide great insights into the multiple aspects of crop evolution, including the identification of crop progenitors (Burke *et al.*, 2007; Konishi *et al.*, 2006; Li *et al.*, 2006), the localization and timing of domestication events, and the demographics of domestication (Liu and Burke, 2006).

Several genes that were critical for domestication or crop improvement have been identified, influencing fruit size (Cong *et al.*, 2008; Frary *et al.*, 2000; Nesbitt and Tanksley, 2002) and shape (Fan *et al.*, 2006), seed dispersal (Konishi *et al.*, 2006), tillering (Li *et al.*, 2003), seed color (Brooks *et al.*, 2008; Konishi *et al.*, 2008; Sweeney *et al.*, 2007), and many other traits (Tang *et al.*, 2010) (Table 3.2). Mutagenic events in such genes, which were naturally present in the progenitor plant population, were critical for the domestication of crop plants. Several mutagenic events that were part of the domestication process in the rice crop are illustrated in Fig. 3.4. When present in the wild-type progenitor plants, such mutated alleles were presumed to reduce reproductive fitness. Hence, such mutated alleles should occur in the wild progenitors only at relatively low frequencies (Tang *et al.*, 2010). Not all the "domestication alleles" are recessive or loss-of-function alleles, whose persistence can be easily understandable; an appreciable number of such alleles are additive or dominant and which have a modified function (Doebley *et al.*, 2006).

High-throughput sequencing technologies are only in their early phases with regard to the crop domestication studies. A recent study utilizing resequencing microarrays to map genome-wide SNP variations revealed phylogenetic relationships, population structure, and introgression history among 20 rice cultivars and landraces (Doebley *et al.*, 2006). Another study utilized short-read sequencing technology to map the "Green Revolution" gene *sd1* (Peng *et al.*, 1999) in a 160-individual recombination inbred population (Huang *et al.*, 2009). A significant number of the domestication alleles associated with the domestication traits are yet to be identified. Recent advances in the exome sequencing technology can be particularly useful in fast-pace identification of the domestication alleles and any mutations/sequence alterations in such alleles that would have favored domestication of the desired traits.

D. Exome Sequencing for the Aid of Traditional Breeders

Traditional plant breeding has been phenomenal in producing crop varieties of desirable traits. High yielding and disease-resistant varieties represent the

Table 3.2 Mapped domestication-related alterations in genes

Crop	Trait	Gene	Gene function	Molecular change	Reference
Maize	Plant and inflorescence structure	*teosinte branched1 (tb1)*	TCP domain-containing transcription factor	SNPs in 60–90 kb 5' to the tb1 transcribed sequence	Clark *et al.* (2004)
Maize	Seed casing (glume hardness, size, and curvature)	*teosinte glume architecture1 (tga1)*	SBP domain containing transcription factor	5 SNPs in the 5'-UTR and 1 SNP in the coding region	Wang *et al.* (2005)
Rice	Grain filling	*grain incomplete filling 1 (GIF1), OsCIN1*	Cell wall invertase	Segmental duplication	Wang *et al.* (2008, 2010a), Xue *et al.*(2008)
Rice	Flowering time	*Ghd7*	CCT (CO, CO-LIKE and TIMING OF CAB1) domain protein	SNPs insertion/ deletions in coding sequences	Xue *et al.* (2008)
Rice	Grains number per panicle	*Gn1a*	Cytokinin oxidase/ dehydrogenase	16 bp deletion in ORF causing frameshift	Ashikari *et al.* (2005)
Rice	Seed width	*qSW5*	Not known	1212-bp deletion and several SNPs	Shomura *et al.* (2008)
Rice	Seed shattering	*qSH1, sh4*	Transcription factors	SNPs in promoter and coding regions, respectively	Ishikawa *et al.* 2010), Konishi *et al.* (2008), Li *et al.* (2006), Thurber *et al.* (2010)

Rice	Plant architecture, including tiller angle and number of tillers	*Prostrate growth 1 (PROG1)*	Zn-finger transcription factor	SNP in coding region	Jin *et al.* (2008), Tan *et al.* (2008)
Rice	Seed pericarp color	*Rc, Rd*	bHLH Transcription factor, dihydroflavinol-4-reductase, respectively	Deletions or SNPs causing premature stop codons, respectively	Brooks *et al.* (2008) Konishi *et al.* (2008), Sweeney *et al.*(2007)
Rice	Texture of cooked rice	*Waxy*	Granule-bound rice synthase	SNP at the 5' splice site of the first intron	Hirano *et al.* (1998), Olsen *et al.* (2006), Yamanaka *et al.* (2004)
Tomato	Fruit weight (through controlling carpel cell number)	*ORFX* (*fw2.2 QTL*)	Negative regulator of cell division	SNPs in the promoter causing transcriptional changes	Frary *et al.* (2000), Nesbitt and Tanksley (2002)
Tomato	Fruit weight (through organ number determination)	*fasciated*	YABBY-like transcription factor	Seven 6- to 8-kb insertion in the first intron	Cong *et al.* (2008)
Wheat	Inflorescence structure	*Q*	Transcriptional regulator (AP2)	SNPs in the coding sequence	Faris and Gill (2002), Simons *et al.* (2006)

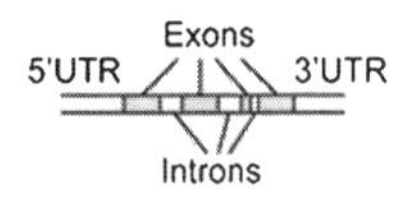

1. Mutation in 5' UTR

Example: Glabrous leaf and hull mutants in rice by SNP in 5' UTR of *gl1*
Detection method: Genomic sequencing/RNA-Seq

2. Mutation in 3' UTR

Example: Cytoplasmic male sterility by truncation in 3' UTR of *rps3* in sugar beet
Detection method: Genomic sequencing/RNA-Seq

3. Mutations in exonic regions

Example: Seed shattering in rice by a SNP in the coding region in the gene *sh4*
Detection method: Genome Seq/ Exome Seq/ RNA-Seq

4. Mutations in splice sites

Example: Glutinous texture in rice by a SNP in splice donar site in the gene *waxy*
Detection method: Genome Seq/ Exome Seq/ RNA-Seq

5. Intragenic deletions/truncations

Example: Seed pericarp color change in rice by 14-bp deletion in *Rc* gene leading to truncation
Detection method: Genome Seq/ Exome Seq/ RNA-Seq

6. Gene duplication

Example: Changes in tomato fruit shape by LTR-mediated duplication of *SUN* gene
Detection method: Genome Sequencing

7. Changes in gene expression

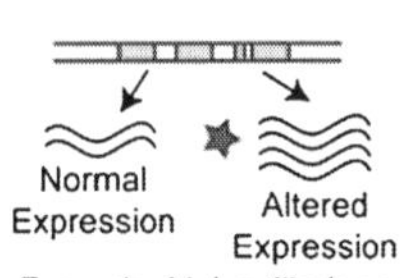

Example: Maize tillering gene (*tb1*)
Detection method: Expression microarray/ RNA-Seq

8. Pseudogene

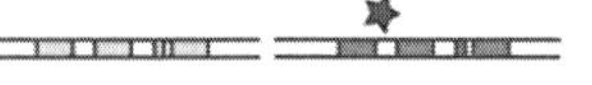

Example: *IPS1* serves as a pseudogene to *PHO2* and sequesters miR-399
Detection method: GenomeSeq/RNA-Seq

Figure 3.4 ***Utility of next-generation sequencing methods in exploring various types of genetic changes involved in crop domestication.*** The illustrations indicate genetic alterations as shown by red stars on representative region of gene that is transcribed. The yellow regions indicate exons. One example and potential methods of detection are given for each type of genetic alterations. For interpretation of the references to color in this figure legend, the reader is referred to the online version of this book.

greatest triumphs of traditional plant breeding in improving food stability and stabilizing supplies of fiber and vegetable oils. Such improvements have been carried out by pedigree method, bulk method, or backcross methods of breeding. The success of such breeding strategies depends on identifying the high yield or disease-resistant genes and their sources. Locating a satisfactory source for such disease-resistant genes by continued observations and tests, as already performed by traditional breeders and pathologists on most crop plants, is not a huge challenge. However, tracing the alleles and the corresponding mutations can be a daunting task. Furthermore, a high yield trait in most cases seems to be an outcome of multiple physiological and morphological traits, and hence, is governed by multiple genes or quantitative trait loci (QTLs). Some such traits for the rice crop include: high stomatal conductance, large rachis-branch number per primary rachis branch, large spikelet number on secondary rachis branches, intermediate tillering ability, high and moderate leaf N contents before and after the late reproductive stage, respectively, high nonstructural carbohydrate accumulation at the onset of grain filling. For such complex traits with poor genetic understanding, breeders tend to shift the mean multi-trait response in a particular direction with an overall favorable outcome.

As discussed in the previous section, the next generation sequencing strategies can help identify the genetic sources for such traits. More importantly, exome sequencing can serve as an affordable method to identify and test the presence of such QTLs or alleles. Also, exome sequencing can serve as a unique affordable tool to test for the inheritance of such genetic markers for routine testing in the cultivars under development. Sequence-based preevaluation of the newly developed varieties can be a time saver in comparison to the large-scale field trials for the phenotypes. Exome sequencing can be monumental in providing large-scale genetic resources for the crop improvement. Future whole-exome sequence analyses or individual allele/QTL sequence analyses of the varieties under development can be compared to the original database(s) to investigate the genetic transmittance of desirable traits.

E. Exome Sequencing and Symbiotic Cropping Strategy Management

Identifying the optimal growth conditions is very critical for the best yields. Traditionally, the growth conditions are optimized by testing the crop under all the given conditions. Knowing the exact genetic makeup and genomic architecture of crop plants can help predict the optimal growth conditions. Several crop plants, which benefit from symbiotic or other coexistential ecosystems, can be genetically engineered to efficiently utilize natural or artificial ecosystems. Several researchers have mathematically or even experimentally designed artificial ecosystems. Hu *et al.* (2010) have utilized coherent experimental data and mathematical simulation to demonstrate that different antibiotics levels and initial cell densities can result in correlated population dynamics such as extinction, obligatory mutualism, facultative mutualism, and commensalism. Kambam *et al.* (2008) designed and analyzed a synthetic symbiotic ecosystem with explicitly defined genetic nature of the microbial interactions. Exome sequencing can help identify the genes and establish the presence of functional gene sets that are involved in symbiotic or other coexistential systems. Such information can be pertinent to planning the crop management and improvement strategies that utilize preexisting and newly developed coexistential systems.

IV. CROP IMPROVEMENT BEYOND EXOME SEQUENCING

Exome sequencing presents a very practical method of assessing genomic alterations in the exon sequences and hence can be a very handy tool in generating resources for crop improvement. However, exome sequencing lacks on several critical issues and some of them are described below:

A. Whole-Genome Sequencing

Whole-genome sequencing, as discussed previously, presents a potential tool for mapping the entire genome and hence is the most exhaustive way of studying genetic transmission of alleles/QTLs. Several aspects of the genome, which are left uncovered by the exomic sequencing, can be explored by whole-genome sequencing. Some of them are discussed below.

1. MicroRNAs (miRNAs) and other noncoding RNAs

Small RNAs (sRNAs), including microRNAs (miRNAs), short interfering RNAs (siRNAs), and transacting siRNAs (ta-siRNAs), are crucial components of epigenetic processes and gene networks involved in development and homeostasis (Ruiz-Ferrer and Voinnet, 2009; Zhang *et al.*, 2011b). Endogenous noncoding small RNAs (21–24 nt) and miRNAs, which are noncoding and are not included in the exome, present an important regulatory mechanism for fine-tuning the expression of genes. Studies have indicated that noncoding small RNA loci, like protein-coding genes, could be targets of domestication selection and play an important role in crop (e.g., rice) domestication, and improvement (Wang *et al.*, 2010c). Furthermore, miRNA expression has been correlated with host–pathogen interactions. Diverse sets of wheat long nonprotein coding RNAs (npcRNAs) have been identified that are responsive to powdery mildew infection and heat stress and could function in wheat responses to both biotic and abiotic stresses (Xin *et al.*, 2011). Nitrogen (N) starvation responses are regulated by miRNA169, which is substantially downregulated in both roots and shoots by N starvation, and the transgenic *Arabidopsis* plants overexpressing miR169a accumulate less N and are more sensitive to N stress than the wild type (Zhao *et al.*, 2011). miRNAs also regulate development and reproduction in plants. Transgenic tomato plants overexpressing sly-miR156a demonstrate changes in multiple vegetative and reproductive traits (Zhang *et al.*, 2011b). Stage- and tissue-specific modulation of multiple miRNAs and their targets has also been identified during somatic embryogenesis of Valencia sweet orange (Wu et al., 2011). miRNAs also play key roles in plant acclimatization. Overexpression of miR169 confers drought resistance in transgenic tomato plants (Zhang *et al.*, 2011a). In short, miRNAs play essential roles in several physiological processes and can be genetically manipulated. However, since the miRNAs are mostly coded by nonexonic regions, exome sequencing will fail to provide any information on sequence alterations or expression levels of miRNAs. The sequence alterations in miRNA or other small noncoding RNAs can be mapped by genomic sequencing or RNA-Seq, while the expression levels can be determined by RNA-Seq or expression microarrays.

2. 3′- and 5′-UTR regions

The untranslated regions (UTRs) in mRNA play critical role of regulating the stability, function, and localization of mRNA. The 3'-UTRs of mRNA also serve as templates for miRNA binding that regulates the turnover and/or function of the mRNA. The rice storage protein *glutelin B-1 (GluB-1)* 3'-UTR functions as a faithful terminator and that termination at the specific sites may play an important role in mRNA stability and/or translatability, resulting in higher levels of protein accumulation (Yang *et al.*, 2009). Rice *glutelin* RNAs are localized to the cisternal endoplasmic reticulum (ER) by a regulated RNA transport process requiring specific *cis*-localization elements, two of which are located at the 5' and 3' ends of the coding sequences and a third one is located within the 3'-UTR (Washida *et al.*, 2009). Similarly, 5'-UTRs are also involved in the stabilization of mRNA and the regulation of gene expression. In order to characterize the glabrous leaf and hull mutants (*gl1*) of rice, Li *et al.* (2010a) cloned and sequenced the target region from four *gl1* mutants (*gl1-1*, *gl1-2*, *gl1-3*, and *gl1-4*) and four glabrous rice varieties (Jackson, Jefferson, Katy, and Lemont), all of which demonstrated the same single point mutation (A → T) that occurred in the 5'-UTR of the locus Os05g0118900 (corresponding to the 3'-UTR of *STAR2*). The 5'-UTRs of *Glb-1* and *GluA-1* enhanced the expression levels of β-glucuronidase (GUS) reporter gene driven by the *GluC* promoter for the rice-seed storage-protein glutelin in stable transgenic rice lines or in transient expression protoplasts (Liu *et al.*, 2010). Similarly, 5'-UTR of the rice polyubiquitin gene (*rubi3*) can enhance transcription in a tissue-specific manner in transgenic rice plants (Lu *et al.*, 2008). Furthermore, the 5'-UTR of the rice alcohol dehydrogenase gene serves as a translational enhancer in both dicotyledonous and monocotyledonous plant cells (Sugio *et al.*, 2008).

3. Pseudogenes

Pseudogenes are DNA sequences that have high homology to known functional genes but are not coded into proteins. Since they are not coded into proteins, they are largely excluded from exome sequencing. For long time, the pseudogenes are thought to serve no function. However, recent studies indicate that the pseudogenes also code for mRNA, which although is not translated, may serve as a scaffold for miRNAs. The 3'-UTRs of pseudogene mRNA may bind to similar miRNA as does the 3'-UTR of the mRNA from a functional gene, depending on the sequence homology. However, since the pseudogene mRNA is not coded into a protein, it only serves to decrease the actual concentration of miRNA available to bind the 3'-UTR of the mRNA from the functional gene and thereby modulate the expression of the functional gene (Ebert and Sharp, 2010). Hence, any sequence variations in the 3'-UTRs of the pseudogenes will have serious implications on the expression of functional genes and should be studied with care. One such example of target mimicry has been demonstrated for nonprotein coding gene *IPS1* (INDUCED BY PHOSPHATE STARVATION1) from *Arabidopsis thaliana* (Franco-Zorrilla *et al.*, 2007). *IPS1* contains a motif with partial sequence complementarity to the phosphate (P_i) starvation-induced miRNA miR-399. Hence, *IPS1* RNA sequesters miR-399, without itself getting cleaved, and results in increased accumulation of the miR-399 target *PHO2* mRNA that leads to reduction in shoot P_i content.

B. Transcriptomics

Transcriptome consists of all the mRNA that is transcribed at a given point in a given tissue. Unlike the genome or exome, which are fixed for all the cells of an organism, the transcriptome is highly variable and depends upon the environment. The transcriptome can be analyzed either at the level of the sequence of RNA by next-generation sequencing or the expression of RNA.

1. RNA-Seq

RNA-Seq is the sequencing approach that utilizes next-generation sequencing technology to study the entire transcriptome. RNA-Seq is a high-throughput alternative to the traditional RNA/cDNA cloning and sequencing strategies. Furthermore, RNA-Seq also provides information on the expression levels of the transcripts and the alternate splice variants. Sudini *et al.* (2011) explored soil bacterial communities in different peanut cropping sequences using molecular approaches such as Ribosomal Intergenic Spacer Analysis (RISA) combined with 16S rRNA cloning and sequencing. Genomic sequencing or RNA-Seq can replace these techniques and provide even more information on the bacterial communities. RNA-Seq has been applied to sequence the transcriptomes of several crop plants including, soybean, rice, and grapes (Lu *et al.*, 2010; Severin *et al.*, 2010; Zenoni *et al.*, 2010; Zhang *et al.*, 2010). However, exome sequencing will be of limited importance for such analysis. Host–pathogen interactions and drought and salinity stress responses have also been successfully characterized by utilizing RNA-Seq (Deyholos, 2010; DiGuistini *et al.*, 2011).

2. Expression microarrays

The effect of the environment on the mRNA expression profiles in a tissue-specific manner can be obtained by expression microarrays. Although, not completely indicative of function, mRNA expression profiles can accurately predict how a given genomic architecture will respond to the environment. Significant progress has been made in understanding the expression patterns in response to the environmental cues. Gene expression (*OsCESA* gene) patterns have been mapped through the entire life cycle of rice by performing microarray analysis (Wang *et al.*, 2010b). Gene expression profiling of entire gene families (e.g., Arabinogalactan, *OsBC1L*, *Hsp20* gene families in rice) have been mapped through microarray analyses (Dai *et al.*, 2009; Ma and Zhao, 2010; Ouyang *et al.*, 2009). The key utility of the expression arrays lies in mapping the expression profiles in response to the environmental cues such as nitrogen and phosphorus deprivation, hormonal stimulation, pathogenic stimulations, and circadian rhythms (Qiu *et al.*, 2009). The applications of expression profiling can be endless, however, it can only complement, not replace, the exome studies. In comparison to RNA-Seq, microarrays are limited to what is available on the chip. Furthermore, RNA-Seq studies have almost unlimited dynamic range. However, unlike the expression microarrays, the RNA-Seq studies suffer from the lack of a background to be subtracted from the signal.

C. Metabolomics

Metabolite levels represent final outcome of interactions between genetic and environmental factors and may serve as an indicator of plant health under conditions of stress. Genes that regulate biotic stress in plants also regulate the levels of metabolites, e.g., *YK1* gene, whose overexpression in rice leaves prevents infection by brown stripe pathogen, also increases the total amounts of NAD(P)H, which serve as essential cofactors for several enzymes, and tricarboxylic acid (TCA) cycle metabolites such as *cis*-aconitate, isocitrate, and 2-oxoglutarate. Simultaneously, it decreases the levels of fructose-1,6-bisphosphate, glyceraldehydes-3-phosphate, and isocitrate (Hayashi *et al.*, 2005; Takahashi *et al.*, 2006). Alterations in the levels metabolites can be easily quantified by nuclear magnetic resonance (NMR) or mass spectrometry (MS). Mass spectrometric techniques such as Fourier-transform ion cyclotron (FT-MS) or capillary electrophoresis-electrospray ionization (CE/MS) have been successfully applied to identify metabolite levels in plants. Metabolic alterations in *Arabidopsis* in response to several environmental triggers have been extensively investigated by metabolomic studies (Fukushima *et al.*, 2011; Kusano *et al.*, 2011). Although exome sequencing can at least partially predict the behavior of crop plant in response to several environmental cues, it cannot accurately measure the outcome. Metabolomic studies cannot only measure the response of plant triggered by the environmental cues but they can also measure the differential response in different plant parts, a task not feasible by the sequencing studies of any sort (Sanchez *et al.*, 2011). However, metabolite levels are indeed an outcome of genetic differences and metabolic differences in cooked rice of multiple varieties have been shown to correlate very well with SNPs in the coding regions that can be measured by the exome sequencing (Heuberger *et al.*, 2011).

D. Proteomics

Similar to metabolome, proteome represents an outcome of interaction between genetic and environmental factors that regulate gene expression.

Mass spectrometric proteomic profiling has been extensively utilized to study the changes in proteomes as a result of biotic and abiotic stresses in crop plants. Proteomics analysis has been utilized to identify the proteins that play a critical role in regulating adaptation activities following exposure to salt (NaCl) stress in order to facilitate ion homeostasis in cucumber seedling roots (Du *et al.*, 2010). Similarly, proteomic analyses have been performed to identify salt stress-induced genes in other crops such as rice (Guo and Song, 2009; Li *et al.*, 2010b). Comparative proteomics has also been utilized to identify molecular mechanisms involved in physiological processes such as the role of oxidative stress on carotenogensis in sweet orange (Pan *et al.*, 2009). Proteomics has also provided significant insights into plant–pathogen interactions. Acclimation to osmotic stress has also been studied by proteomic analyses (Carpentier *et al.*, 2007). Matrix-assisted laser desorption/ionization time-of-flight (MALDI-TOF) MS and tandem TOF/TOF MS analyses have led to identification of novel proteins and hence, genes regulating the soybean defense mechanism against Soybean mosaic virus (SMV) and several other plant–pathogen interactions (Gonzalez-Fernandez *et al.*, 2010; Nagy and Pogany, 2010; Seidl *et al.*, 2011; Yang *et al.*, 2010). Plant growth and development are also influenced by light-regulated diurnal rhythms as well as endogenous clock-regulated circadian rhythms. Proteomic analyses have also provided better insights into light-regulated and clock-regulated rhythms (Hwang *et al.*, 2011). Comparative proteomic analyses have also shed light on the symbiotic processes such as root nodulation in soybean (Lim *et al.*, 2010). Proteomic analyses have also been helpful in studying molecular mechanisms of several other biological processes in response to hormone treatments (Martinez-Esteso *et al.*, 2011; Shan *et al.*, 2011; Wienkoop *et al.*, 2010). However, proteomic analyses without genomic analyses are of limited importance and require more complex analyses in predicting trait transmission or domestication.

V. CONCLUSIONS

Exome sequencing presents an excellent and practical (cost and time efficient) tool for crop improvement. Knowing the exact genetic makeup of the crop variety can help manage the crop more efficiently. However, exome sequencing by itself will not suffice for efficient crop improvement; rather, an integrative approach utilizing multiple omics and bioinformatics platforms and integration of their outcomes will serve as an efficient crop improvement strategy.

COMPETING INTERESTS

No existing financial or non-financial competing interests.

ACKNOWLEDGMENTS

We would like thank Ms. Preeti Sirohi (student, M.Tech., Biotech) for help with editing the manuscript.

ABBREVIATIONS

UTR	Untranslated region
QTL	Quantitative trait loci
PCR	Polymerase chain reaction
CBOL	Consortium for the Barcode of Life
sRNAs	Small RNAs
npcRNAs	Nonprotein-coding RNAs
miRNAs	microRNAs
siRNAs	Short interfering RNAs
ta-siRNAs	transacting siRNAs
MS	Mass spectrometry
NMR	Nuclear magnetic resonance
MALDI-TOF	Matrix assisted laser desorption/ionization time-of-flight
FT	Fourier transform
CE	Capillary electrophoresis
RISA	Ribosomal Intergenic Spacer Analysis

REFERENCES

Ashikari, M., Sakakibara, H., Lin, S., Yamamoto, T., Takashi, T., Nishimura, A., Angeles, E.R., Qian, Q., Kitano, H., Matsuoka, M., 2005. Cytokinin oxidase regulates rice grain production. Science 309, 741–745.

Brooks, S.A., Yan, W., Jackson, A.K., Deren, C.W., 2008. A natural mutation in rc reverts white-rice-pericarp to red and results in a new, dominant, wild-type allele: Rc-g. Theor. Appl. Genet. 117, 575–580.

Burke, J.M., Burger, J.C., Chapman, M.A., 2007. Crop evolution: from genetics to genomics. Curr. Opin. Genet. Dev. 17, 525–532.

Calvitti, M., Moretti, R., Lampazzi, E., Bellini, R., Dobson, S.L., 2010. Characterization of a new Aedes albopictus (Diptera: Culicidae)-Wolbachia pipientis (Rickettsiales: Rickettsiaceae) symbiotic association generated by artificial transfer of the wPip strain from Culex pipiens (Diptera: Culicidae). J. Med. Entomol. 47, 179–187.

Carpentier, S.C., Witters, E., Laukens, K., Van Onckelen, H., Swennen, R., Panis, B., 2007. Banana (Musa spp.) as a model to study the meristem proteome: acclimation to osmotic stress. Proteomics 7, 92–105.
Carroll, R.K., Shelburne 3rd, S.A., Olsen, R.J., Suber, B., Sahasrabhojane, P., Kumaraswami, M., Beres, S.B., Shea, P.R., Flores, A.R., Musser, J.M., 2011. Naturally occurring single amino acid replacements in a regulatory protein alter streptococcal gene expression and virulence in mice. J. Clin. Invest. 121, 1956–1968.
Clark, R.M., Linton, E., Messing, J., Doebley, J.F., 2004. Pattern of diversity in the genomic region near the maize domestication gene tb1. Proc. Natl. Acad. Sci. USA 101, 700–707.
Cong, B., Barrero, L.S., Tanksley, S.D., 2008. Regulatory change in YABBY-like transcription factor led to evolution of extreme fruit size during tomato domestication. Nat. Genet. 40, 800–804.
Dai, X., You, C., Wang, L., Chen, G., Zhang, Q., Wu, C., 2009. Molecular characterization, expression pattern, and function analysis of the OsBC1L family in rice. Plant Mol. Biol. 71, 469–481.
Deyholos, M.K., 2010. Making the most of drought and salinity transcriptomics. Plant Cell Environ. 33, 648–654.
DiGuistini, S., Wang, Y., Liao, N.Y., Taylor, G., Tanguay, P., Feau, N., Henrissat, B., Chan, S.K., Hesse-Orce, U., Alamouti, S.M., Tsui, C.K., Docking, R.T., Levasseur, A., Haridas, S., Robertson, G., Birol, I., Holt, R.A., Marra, M.A., Hamelin, R.C., Hirst, M., Jones, S.J., Bohlmann, J., Breuil, C., 2011. Genome and transcriptome analyses of the mountain pine beetle-fungal symbiont Grosmannia clavigera, a lodgepole pine pathogen. Proc. Natl. Acad. Sci. USA 108, 2504–2509.
Doebley, J.F., Gaut, B.S., Smith, B.D., 2006. The molecular genetics of crop domestication. Cell 127, 1309–1321.
Drmanac, R., Sparks, A.B., Callow, M.J., Halpern, A.L., Burns, N.L., Kermani, B.G., Carnevali, P., Nazarenko, I., Nilsen, G.B., Yeung, G., Dahl, F., Fernandez, A., Staker, B., Pant, K.P., Baccash, J., Borcherding, A.P., Brownley, A., Cedeno, R., Chen, L., Chernikoff, D., Cheung, A., Chirita, R., Curson, B., Ebert, J.C., Hacker, C.R., Hartlage, R., Hauser, B., Huang, S., Jiang, Y., Karpinchyk, V., Koenig, M., Kong, C., Landers, T., Le, C., Liu, J., McBride, C.E., Morenzoni, M., Morey, R.E., Mutch, K., Perazich, H., Perry, K., Peters, B.A., Peterson, J., Pethiyagoda, C.L., Pothuraju, K., Richter, C., Rosenbaum, A.M., Roy, S., Shafto, J., Sharanhovich, U., Shannon, K.W., Sheppy, C.G., Sun, M., Thakuria, J.V., Tran, A., Vu, D., Zaranek, A.W., Wu, X., Drmanac, S., Oliphant, A.R., Banyai, W.C., Martin, B., Ballinger, D.G., Church, G.M., Reid, C.A., 2010. Human genome sequencing using unchained base reads on self-assembling DNA nanoarrays. Science 327, 78–81.
Du, C.X., Fan, H.F., Guo, S.R., Tezuka, T., Li, J., 2010. Proteomic analysis of cucumber seedling roots subjected to salt stress. Phytochemistry 71, 1450–1459.
Ebert, M.S., Sharp, P.A., 2010. Emerging roles for natural microRNA sponges. Curr. Biol. 20, R858–R861.
Edwards, A., Voss, H., Rice, P., Civitello, A., Stegemann, J., Schwager, C., Zimmermann, J., Erfle, H., Caskey, C.T., Ansorge, W., 1990. Automated DNA sequencing of the human HPRT locus. Genomics 6, 593–608.
Eid, J., Fehr, A., Gray, J., Luong, K., Lyle, J., Otto, G., Peluso, P., Rank, D., Baybayan, P., Bettman, B., Bibillo, A., Bjornson, K., Chaudhuri, B., Christians, F., Cicero, R., Clark, S., Dalal, R., Dewinter, A., Dixon, J., Foquet, M., Gaertner, A., Hardenbol, P., Heiner, C., Hester, K., Holden, D., Kearns, G., Kong, X., Kuse, R., Lacroix, Y., Lin, S., Lundquist, P., Ma, C., Marks, P., Maxham, M., Murphy, D., Park, I., Pham, T., Phillips, M., Roy, J., Sebra, R., Shen, G., Sorenson, J., Tomaney, A., Travers, K., Trulson, M., Vieceli, J., Wegener, J., Wu, D., Yang, A., Zaccarin, D., Zhao, P.,

Zhong, F., Korlach, J., Turner, S., 2009. Real-time DNA sequencing from single polymerase molecules. Science 323, 133–138.
Fan, C., Xing, Y., Mao, H., Lu, T., Han, B., Xu, C., Li, X., Zhang, Q., 2006. GS3, a major QTL for grain length and weight and minor QTL for grain width and thickness in rice, encodes a putative transmembrane protein. Theor. Appl. Genet. 112, 1164–1171.
Faris, J.D., Gill, B.S., 2002. Genomic targeting and high-resolution mapping of the domestication gene Q in wheat. Genome 45, 706–718.
Flusberg, B.A., Webster, D.R., Lee, J.H., Travers, K.J., Olivares, E.C., Clark, T.A., Korlach, J., Turner, S.W., 2010. Direct detection of DNA methylation during single-molecule, real-time sequencing. Nat. Methods 7, 461–465.
Franco-Zorrilla, J.M., Valli, A., Todesco, M., Mateos, I., Puga, M.I., Rubio-Somoza, I., Leyva, A., Weigel, D., Garcia, J.A., Paz-Ares, J., 2007. Target mimicry provides a new mechanism for regulation of microRNA activity. Nat. Genet. 39, 1033–1037.
Frary, A., Nesbitt, T.C., Grandillo, S., Knaap, E., Cong, B., Liu, J., Meller, J., Elber, R., Alpert, K.B., Tanksley, S.D., 2000. fw2.2: a quantitative trait locus key to the evolution of tomato fruit size. Science 289, 85–88.
Fukushima, A., Kusano, M., Redestig, H., Arita, M., Saito, K., 2011. Metabolomic correlation-network modules in Arabidopsis based on a graph-clustering approach. BMC Syst. Biol. 5, 1.
Gonzalez-Fernandez, R., Prats, E., Jorrin-Novo, J.V., 2010. Proteomics of plant pathogenic fungi. J. Biomed. Biotechnol. 2010, 932527.
Green, R.E., Krause, J., Briggs, A.W., Maricic, T., Stenzel, U., Kircher, M., Patterson, N., Li, H., Zhai, W., Fritz, M.H., Hansen, N.F., Durand, E.Y., Malaspinas, A.S., Jensen, J.D., Marques-Bonet, T., Alkan, C., Prufer, K., Meyer, M., Burbano, H.A., Good, J.M., Schultz, R., Aximu-Petri, A., Butthof, A., Hober, B., Hoffner, B., Siegemund, M., Weihmann, A., Nusbaum, C., Lander, E.S., Russ, C., Novod, N., Affourtit, J., Egholm, M., Verna, C., Rudan, P., Brajkovic, D., Kucan, Z., Gusic, I., Doronichev, V.B., Golovanova, L.V., Lalueza-Fox, C., de la Rasilla, M., Fortea, J., Rosas, A., Schmitz, R.W., Johnson, P.L., Eichler, E.E., Falush, D., Birney, E., Mullikin, J.C., Slatkin, M., Nielsen, R., Kelso, J., Lachmann, M., Reich, D., Paabo, S., 2010. A draft sequence of the Neandertal genome. Science 328, 710–722.
Greenleaf, W.J., Block, S.M., 2006. Single-molecule, motion-based DNA sequencing using RNA polymerase. Science 313, 801.
Guo, Y., Song, Y., 2009. Differential proteomic analysis of apoplastic proteins during initial phase of salt stress in rice. Plant Signal Behav. 4, 121–122.
Hayashi, M., Takahashi, H., Tamura, K., Huang, J., Yu, L.H., Kawai-Yamada, M., Tezuka, T., Uchimiya, H., 2005. Enhanced dihydroflavonol-4-reductase activity and NAD homeostasis leading to cell death tolerance in transgenic rice. Proc. Natl. Acad. Sci. USA 102, 7020–7025.
Hebert, P.D., Ratnasingham, S., deWaard, J.R., 2003. Barcoding animal life: cytochrome c oxidase subunit 1 divergences among closely related species. Proc. Biol. Sci. 270 (Suppl. 1), S96–S99.
Heuberger, A.L., Lewis, M.R., Chen, M.H., Brick, M.A., Leach, J.E., Ryan, E.P., 2011. Metabolomic and functional genomic analyses reveal varietal differences in bioactive compounds of cooked rice. PLoS One 5, e12915.
Hirano, H.Y., Eiguchi, M., Sano, Y., 1998. A single base change altered the regulation of the Waxy gene at the posttranscriptional level during the domestication of rice. Mol. Biol. Evol. 15, 978–987.
Hu, B., Du, J., Zou, R.Y., Yuan, Y.J., 2010. An environment-sensitive synthetic microbial ecosystem. PLoS One 5, e10619.

Huang, X., Feng, Q., Qian, Q., Zhao, Q., Wang, L., Wang, A., Guan, J., Fan, D., Weng, Q., Huang, T., Dong, G., Sang, T., Han, B., 2009. High-throughput genotyping by whole-genome resequencing. Genome Res. 19, 1068–1076.

Hwang, H., Cho, M.H., Hahn, B.S., Lim, H., Kwon, Y.K., Hahn, T.R., Bhoo, S.H., 2011. Proteomic identification of rhythmic proteins in rice seedlings. Biochim. Biophys. Acta 1814, 470–479.

Ishikawa, R., Thanh, P.T., Nimura, N., Htun, T.M., Yamasaki, M., Ishii, T., 2010. Allelic interaction at seed-shattering loci in the genetic backgrounds of wild and cultivated rice species. Genes Genet. Syst. 85, 265–271.

Jin, J., Huang, W., Gao, J.P., Yang, J., Shi, M., Zhu, M.Z., Luo, D., Lin, H.X., 2008. Genetic control of rice plant architecture under domestication. Nat. Genet. 40, 1365–1369.

Kambam, P.K., Henson, M.A., Sun, L., 2008. Design and mathematical modelling of a synthetic symbiotic ecosystem. IET Syst. Biol. 2, 33–38.

Konishi, S., Ebana, K., Izawa, T., 2008. Inference of the japonica rice domestication process from the distribution of six functional nucleotide polymorphisms of domestication-related genes in various landraces and modern cultivars. Plant Cell Physiol. 49, 1283–1293.

Konishi, S., Izawa, T., Lin, S.Y., Ebana, K., Fukuta, Y., Sasaki, T., Yano, M., 2006. An SNP caused loss of seed shattering during rice domestication. Science 312, 1392–1396.

Korlach, J., Bjornson, K.P., Chaudhuri, B.P., Cicero, R.L., Flusberg, B.A., Gray, J.J., Holden, D., Saxena, R., Wegener, J., Turner, S.W., 2010. Real-time DNA sequencing from single polymerase molecules. Methods Enzymol. 472, 431–455.

Kusano, M., Tohge, T., Fukushima, A., Kobayashi, M., Hayashi, N., Otsuki, H., Kondou, Y., Goto, H., Kawashima, M., Matsuda, F., Niida, R., Matsui, M., Saito, K., Fernie, A.R., 2011. Metabolomics reveals comprehensive reprogramming involving two independent metabolic responses of Arabidopsis to UV-B light. Plant J. 67, 354–369.

Li, C., Zhou, A., Sang, T., 2006. Rice domestication by reducing shattering. Science 311, 1936–1939.

Li, W., Wu, J., Weng, S., Zhang, D., Zhang, Y., Shi, C., 2010a. Characterization and fine mapping of the glabrous leaf and hull mutants (gl1) in rice (Oryza sativa L.). Plant Cell Rep. 29, 617–627.

Li, X., Qian, Q., Fu, Z., Wang, Y., Xiong, G., Zeng, D., Wang, X., Liu, X., Teng, S., Hiroshi, F., Yuan, M., Luo, D., Han, B., Li, J., 2003. Control of tillering in rice. Nature 422, 618–621.

Li, X.J., Yang, M.F., Chen, H., Qu, L.Q., Chen, F., Shen, S.H., 2010b. Abscisic acid pretreatment enhances salt tolerance of rice seedlings: proteomic evidence. Biochim. Biophys. Acta 1804, 929–940.

Lim, C.W., Park, J.Y., Lee, S.H., Hwang, C.H., 2010. Comparative proteomic analysis of soybean nodulation using a supernodulation mutant, SS2-2. Biosci. Biotechnol. Biochem. 74, 2396–2404.

Lindeberg, M., Myers, C.R., Collmer, A., Schneider, D.J., 2008. Roadmap to new virulence determinants in Pseudomonas syringae: insights from comparative genomics and genome organization. Mol. Plant Microbe. Interact. 21, 685–700.

Liu, A., Burke, J.M., 2006. Patterns of nucleotide diversity in wild and cultivated sunflower. Genetics 173, 321–330.

Liu, W.X., Liu, H.L., Chai, Z.J., Xu, X.P., Song, Y.R., Qu le, Q., 2010. Evaluation of seed storage-protein gene 5' untranslated regions in enhancing gene expression in transgenic rice seed. Theor. Appl. Genet. 121, 1267–1274.

Lu, J., Sivamani, E., Li, X., Qu, R., 2008. Activity of the 5' regulatory regions of the rice polyubiquitin rubi3 gene in transgenic rice plants as analyzed by both GUS and GFP reporter genes. Plant Cell Rep. 27, 1587–1600.

Lu, T., Lu, G., Fan, D., Zhu, C., Li, W., Zhao, Q., Feng, Q., Zhao, Y., Guo, Y., Huang, X., Han, B., 2010. Function annotation of the rice transcriptome at single-nucleotide resolution by RNA-seq. Genome Res. 20, 1238–1249.

Ma, H., Zhao, J., 2010. Genome-wide identification, classification, and expression analysis of the arabinogalactan protein gene family in rice (*Oryza sativa* L.). J. Exp. Bot. 61, 2647–2668.

Margulies, M., Egholm, M., Altman, W.E., Attiya, S., Bader, J.S., Bemben, L.A., Berka, J., Braverman, M.S., Chen, Y.J., Chen, Z., Dewell, S.B., Du, L., Fierro, J.M., Gomes, X.V., Godwin, B.C., He, W., Helgesen, S., Ho, C.H., Irzyk, G.P., Jando, S.C., Alenquer, M.L., Jarvie, T.P., Jirage, K.B., Kim, J.B., Knight, J.R., Lanza, J.R., Leamon, J.H., Lefkowitz, S.M., Lei, M., Li, J., Lohman, K.L., Lu, H., Makhijani, V.B., McDade, K.E., McKenna, M.P., Myers, E.W., Nickerson, E., Nobile, J.R., Plant, R., Puc, B.P., Ronan, M.T., Roth, G.T., Sarkis, G.J., Simons, J.F., Simpson, J.W., Srinivasan, M., Tartaro, K.R., Tomasz, A., Vogt, K.A., Volkmer, G.A., Wang, S.H., Wang, Y., Weiner, M.P., Yu, P., Begley, R.F., Rothberg, J.M., 2005. Genome sequencing in microfabricated high-density picolitre reactors. Nature 437, 376–380.

Martinez-Esteso, M.J., Selles-Marchart, S., Vera-Urbina, J.C., Pedreno, M.A., Bru-Martinez, R., 2011. DIGE analysis of proteome changes accompanying large resveratrol production by grapevine (Vitis vinifera cv. Gamay) cell cultures in response to methyl-beta-cyclodextrin and methyl jasmonate elicitors. J. Proteomics 74, 1421–1436.

Maxam, A.M., Gilbert, W., 1977. A new method for sequencing DNA. Proc. Natl. Acad. Sci. USA 74, 560–564.

Mitsui, J., Fukuda, Y., Azuma, K., Tozaki, H., Ishiura, H., Takahashi, Y., Goto, J., Tsuji, S., 2010. Multiplexed resequencing analysis to identify rare variants in pooled DNA with barcode indexing using next-generation sequencer. J. Hum. Genet. 55, 448–455.

Murphy, K.M., Berg, K.D., Eshleman, J.R., 2005. Sequencing of genomic DNA by combined amplification and cycle sequencing reaction. Clin. Chem. 51, 35–39.

Nagy, P.D., Pogany, J., 2010. Global genomics and proteomics approaches to identify host factors as targets to induce resistance against Tomato bushy stunt virus. Adv. Virus Res. 76, 123–177.

Nesbitt, T.C., Tanksley, S.D., 2002. Comparative sequencing in the genus Lycopersicon. Implications for the evolution of fruit size in the domestication of cultivated tomatoes. Genetics 162, 365–379.

O'Brien, H.E., Thakur, S., Guttman, D.S., 2010. Evolution of plant pathogenesis in Pseudomonas syringae: a genomics perspective. Annu. Rev. Phytopathol. 49, 269–289.

Olsen, K.M., Caicedo, A.L., Polato, N., McClung, A., McCouch, S., Purugganan, M.D., 2006. Selection under domestication: evidence for a sweep in the rice waxy genomic region. Genetics 173, 975–983.

Ouyang, Y., Chen, J., Xie, W., Wang, L., Zhang, Q., 2009. Comprehensive sequence and expression profile analysis of Hsp20 gene family in rice. Plant Mol. Biol. 70, 341–357.

Pan, Z., Liu, Q., Yun, Z., Guan, R., Zeng, W., Xu, Q., Deng, X., 2009. Comparative proteomics of a lycopene-accumulating mutant reveals the important role of oxidative stress on carotenogenesis in sweet orange (Citrus sinensis [L.] osbeck). Proteomics 9, 5455–5470.

Pareek, C.S., Smoczynski, R., Tretyn, A., 2011. Sequencing technologies and genome sequencing. J. Appl. Genet. 52, 413–435.

Peng, J., Richards, D.E., Hartley, N.M., Murphy, G.P., Devos, K.M., Flintham, J.E., Beales, J., Fish, L.J., Worland, A.J., Pelica, F., Sudhakar, D., Christou, P., Snape, J.W., Gale, M.D., Harberd, N.P., 1999. 'Green revolution' genes encode mutant gibberellin response modulators. Nature 400, 256–261.

Porreca, G.J., 2010. Genome sequencing on nanoballs. Nat. Biotechnol. 28, 43–44.

Qiu, X., Xie, W., Lian, X., Zhang, Q., 2009. Molecular analyses of the rice glutamate dehydrogenase gene family and their response to nitrogen and phosphorous deprivation. Plant Cell Rep. 28, 1115–1126.

Rothberg, J.M., Hinz, W., Rearick, T.M., Schultz, J., Mileski, W., Davey, M., Leamon, J.H., Johnson, K., Milgrew, M.J., Edwards, M., Hoon, J., Simons, J.F., Marran, D., Myers, J.W., Davidson, J.F., Branting, A., Nobile, J.R., Puc, B.P., Light, D., Clark, T.A., Huber, M., Branciforte, J.T., Stoner, I.B., Cawley, S.E., Lyons, M., Fu, Y., Homer, N., Sedova, M., Miao, X., Reed, B., Sabina, J., Feierstein, E., Schorn, M., Alanjary, M., Dimalanta, E., Dressman, D., Kasinskas, R., Sokolsky, T., Fidanza, J.A., Namsaraev, E., McKernan, K.J., Williams, A., Roth, G.T., Bustillo, J., 2011. An integrated semiconductor device enabling non-optical genome sequencing. Nature 475, 348–352.

Ruiz-Ferrer, V., Voinnet, O., 2009. Roles of plant small RNAs in biotic stress responses. Annu. Rev. Plant Biol. 60, 485–510.

Sanchez, D.H., Pieckenstain, F.L., Szymanski, J., Erban, A., Bromke, M., Hannah, M.A., Kraemer, U., Kopka, J., Udvardi, M.K., 2011. Comparative functional genomics of salt stress in related model and cultivated plants identifies and overcomes limitations to translational genomics. PLoS One 6, e17094.

Sanger, F., Coulson, A.R., 1975. A rapid method for determining sequences in DNA by primed synthesis with DNA polymerase. J. Mol. Biol. 94, 441–448.

Seidl, M.F., Van den Ackerveken, G., Govers, F., Snel, B., 2011. A domain-centric analysis of oomycete plant pathogen genomes reveals unique protein organization. Plant Physiol. 155, 628–644.

SenGupta, D.J., Cookson, B.T., 2010. SeqSharp: a general approach for improving cycle-sequencing that facilitates a robust one-step combined amplification and sequencing method. J. Mol. Diagn. 12, 272–277.

Severin, A.J., Woody, J.L., Bolon, Y.T., Joseph, B., Diers, B.W., Farmer, A.D., Muehlbauer, G.J., Nelson, R.T., Grant, D., Specht, J.E., Graham, M.A., Cannon, S.B., May, G.D., Vance, C.P., Shoemaker, R.C., 2010. RNA-Seq Atlas of Glycine max: a guide to the soybean transcriptome. BMC Plant Biol. 10, 160.

Shan, X., Wang, J., Chua, L., Jiang, D., Peng, W., Xie, D., 2011. The role of Arabidopsis Rubisco activase in jasmonate-induced leaf senescence. Plant Physiol. 155, 751–764.

Shomura, A., Izawa, T., Ebana, K., Ebitani, T., Kanegae, H., Konishi, S., Yano, M., 2008. Deletion in a gene associated with grain size increased yields during rice domestication. Nat. Genet. 40, 1023–1028.

Simons, K.J., Fellers, J.P., Trick, H.N., Zhang, Z., Tai, Y.S., Gill, B.S., Faris, J.D., 2006. Molecular characterization of the major wheat domestication gene Q. Genetics 172, 547–555.

Smith, L.M., Sanders, J.Z., Kaiser, R.J., Hughes, P., Dodd, C., Connell, C.R., Heiner, C., Kent, S.B., Hood, L.E., 1986. Fluorescence detection in automated DNA sequence analysis. Nature 321, 674–679.

Sudini, H., Liles, M.R., Arias, C.R., Bowen, K.L., Huettel, R.N., 2011. Exploring soil bacterial communities in different peanut-cropping sequences using multiple molecular approaches. Phytopathology 101, 819–827.

Sugio, T., Satoh, J., Matsuura, H., Shinmyo, A., Kato, K., 2008. The 5'-untranslated region of the Oryza sativa alcohol dehydrogenase gene functions as a translational enhancer in monocotyledonous plant cells. J. Biosci. Bioeng. 105, 300–302.

Sweeney, M.T., Thomson, M.J., Cho, Y.G., Park, Y.J., Williamson, S.H., Bustamante, C.D., McCouch, S.R., 2007. Global dissemination of a single mutation conferring white pericarp in rice. PLoS Genet. 3, e133.

Takahashi, H., Hayashi, M., Goto, F., Sato, S., Soga, T., Nishioka, T., Tomita, M., Kawai-Yamada, M., Uchimiya, H., 2006. Evaluation of metabolic alteration in transgenic rice overexpressing dihydroflavonol-4-reductase. Ann. Bot. 98, 819–825.

Tan, L., Li, X., Liu, F., Sun, X., Li, C., Zhu, Z., Fu, Y., Cai, H., Wang, X., Xie, D., Sun, C., 2008. Control of a key transition from prostrate to erect growth in rice domestication. Nat. Genet. 40, 1360–1364.

Tang, H., Sezen, U., Paterson, A.H., 2010. Domestication and plant genomes. Curr. Opin. Plant Biol. 13, 160–166.

Thompson, J.F., Steinmann, K.E., 2010. Single molecule sequencing with a HeliScope genetic analysis system. Curr. Protoc. Mol. Biol. Chapter 7, Unit7 10.

Thurber, C.S., Reagon, M., Gross, B.L., Olsen, K.M., Jia, Y., Caicedo, A.L., 2010. Molecular evolution of shattering loci in U.S. weedy rice. Mol. Ecol. 19, 3271–3284.

Venu, R.C., Zhang, Y., Weaver, B., Carswell, P., Mitchell, T.K., Meyers, B.C., Boehm, M.J., Wang, G.L., 2011. Large scale identification of genes involved in plant-fungal interactions using Illumina's sequencing-by-synthesis technology. Methods Mol. Biol. 722, 167–178.

Wang, E., Wang, J., Zhu, X., Hao, W., Wang, L., Li, Q., Zhang, L., He, W., Lu, B., Lin, H., Ma, H., Zhang, G., He, Z., 2008. Control of rice grain-filling and yield by a gene with a potential signature of domestication. Nat. Genet. 40, 1370–1374.

Wang, E., Xu, X., Zhang, L., Zhang, H., Lin, L., Wang, Q., Li, Q., Ge, S., Lu, B.R., Wang, W., He, Z., 2010a. Duplication and independent selection of cell-wall invertase genes GIF1 and OsCIN1 during rice evolution and domestication. BMC Evol. Biol. 10, 108.

Wang, H., Nussbaum-Wagler, T., Li, B., Zhao, Q., Vigouroux, Y., Faller, M., Bomblies, K., Lukens, L., Doebley, J.F., 2005. The origin of the naked grains of maize. Nature 436, 714–719.

Wang, L., Guo, K., Li, Y., Tu, Y., Hu, H., Wang, B., Cui, X., Peng, L., 2010b. Expression profiling and integrative analysis of the CESA/CSL superfamily in rice. BMC Plant Biol. 10, 282.

Wang, Y., Shen, D., Bo, S., Chen, H., Zheng, J., Zhu, Q.H., Cai, D., Helliwell, C., Fan, L., 2010c. Sequence variation and selection of small RNAs in domesticated rice. BMC Evol. Biol. 10, 119.

Washida, H., Kaneko, S., Crofts, N., Sugino, A., Wang, C., Okita, T.W., 2009. Identification of cis-localization elements that target glutelin RNAs to a specific subdomain of the cortical endoplasmic reticulum in rice endosperm cells. Plant Cell Physiol. 50, 1710–1714.

Wheeler, D.A., Srinivasan, M., Egholm, M., Shen, Y., Chen, L., McGuire, A., He, W., Chen, Y.J., Makhijani, V., Roth, G.T., Gomes, X., Tartaro, K., Niazi, F., Turcotte, C.L., Irzyk, G.P., Lupski, J.R., Chinault, C., Song, X.Z., Liu, Y., Yuan, Y., Nazareth, L., Qin, X., Muzny, D.M., Margulies, M., Weinstock, G.M., Gibbs, R.A., Rothberg, J.M., 2008. The complete genome of an individual by massively parallel DNA sequencing. Nature 452, 872–876.

Wienkoop, S., Baginsky, S., Weckwerth, W., 2010. Arabidopsis thaliana as a model organism for plant proteome research. J. Proteomics 73, 2239–2248.

Wu, X.M., Liu, M.Y., Ge, X.X., Xu, Q., Guo, W.W., 2011. Stage and tissue-specific modulation of ten conserved miRNAs and their targets during somatic embryogenesis of Valencia sweet orange. Planta 233, 495–505.

Xin, M., Wang, Y., Yao, Y., Song, N., Hu, Z., Qin, D., Xie, C., Peng, H., Ni, Z., Sun, Q., 2011. Identification and characterization of wheat long non-protein coding RNAs responsive to powdery mildew infection and heat stress by using microarray analysis and SBS sequencing. BMC Plant Biol. 11, 61.

Xue, W., Xing, Y., Weng, X., Zhao, Y., Tang, W., Wang, L., Zhou, H., Yu, S., Xu, C., Li, X., Zhang, Q., 2008. Natural variation in Ghd7 is an important regulator of heading date and yield potential in rice. Nat. Genet. 40, 761–767.

Yamanaka, S., Nakamura, I., Watanabe, K.N., Sato, Y., 2004. Identification of SNPs in the waxy gene among glutinous rice cultivars and their evolutionary significance during the domestication process of rice. Theor. Appl. Genet. 108, 1200–1204.

Yang, H., Huang, Y., Zhi, H., Yu, D., 2010. Proteomics-based analysis of novel genes involved in response toward Soybean mosaic virus infection. Mol. Biol. Rep. 38, 511–521.

Yang, L., Wakasa, Y., Kawakatsu, T., Takaiwa, F., 2009. The 3'-untranslated region of rice glutelin GluB-1 affects accumulation of heterologous protein in transgenic rice. Biotechnol. Lett. 31, 1625–1631.

Zenoni, S., Ferrarini, A., Giacomelli, E., Xumerle, L., Fasoli, M., Malerba, G., Bellin, D., Pezzotti, M., Delledonne, M., 2010. Characterization of transcriptional complexity during berry development in Vitis vinifera using RNA-Seq. Plant Physiol. 152, 1787–1795.

Zhang, G., Guo, G., Hu, X., Zhang, Y., Li, Q., Li, R., Zhuang, R., Lu, Z., He, Z., Fang, X., Chen, L., Tian, W., Tao, Y., Kristiansen, K., Zhang, X., Li, S., Yang, H., Wang, J., 2010. Deep RNA sequencing at single base-pair resolution reveals high complexity of the rice transcriptome. Genome Res. 20, 646–654.

Zhang, X., Zou, Z., Gong, P., Zhang, J., Ziaf, K., Li, H., Xiao, F., Ye, Z., 2011a. Over-expression of microRNA169 confers enhanced drought tolerance to tomato. Biotechnol. Lett. 33, 403–409.

Zhang, X., Zou, Z., Zhang, J., Zhang, Y., Han, Q., Hu, T., Xu, X., Liu, H., Li, H., Ye, Z., 2011b. Over-expression of sly-miR156a in tomato results in multiple vegetative and reproductive trait alterations and partial phenocopy of the sft mutant. FEBS Lett. 585, 435–439.

Zhao, M., Ding, H., Zhu, J.K., Zhang, F., Li, W.X., 2011. Involvement of miR169 in the nitrogen-starvation responses in Arabidopsis. New Phytol. 190, 906–915.

Applications of Functional Protein Microarrays in Basic and Clinical Research

Heng Zhu[*,†,‡] **and Jiang Qian**[‡,§]

[*]Department of Pharmacology and Molecular Sciences, Johns Hopkins University School of Medicine, Baltimore, MD, USA
[†]The High-Throughput Biology Center, Johns Hopkins University School of Medicine, Baltimore, MD, USA
[‡]The Oncology Center, Johns Hopkins University School of Medicine, Baltimore, MD, USA
[§]Department of Ophthalmology, Johns Hopkins University School of Medicine, Baltimore, MD, USA

Contents

Advances in Genetics, Volume 79
ISSN 0065-2660,
http://dx.doi.org/10.1016/B978-0-12-394395-8.00004-9

Abstract

The protein microarray technology provides a versatile platform for characterization of hundreds of thousands of proteins in a highly parallel and high-throughput manner. It is viewed as a new tool that overcomes the limitation of DNA microarrays. On the basis of its application, protein microarrays fall into two major classes: analytical and functional protein microarrays. In addition, tissue or cell lysates can also be directly spotted on a slide to form the so-called "reverse-phase" protein microarray. In the last decade, applications of functional protein microarrays in particular have flourished in studying protein function and construction of networks and pathways. In this chapter, we will review the recent advancements in the protein microarray technology, followed by presenting a series of examples to illustrate the power and versatility of protein microarrays in both basic and clinical research. As a powerful technology platform, it would not be surprising if protein microarrays will become one of the leading technologies in proteomic and diagnostic fields in the next decade.

I. INTRODUCTION

The concept of microarray technology was first put forward by Ekins (1989)over 20 years ago. An ambient analyte theory was proposed that a tiny spot of a purified antibody or protein provides substantially better sensitivity than when used in conventional immunoassay formats as miniaturized features can dramatically enhance detection sensitivity. Though not exactly the same, DNA microarray technology became the first application of this theory and has been tremendously successful in gene expression profiling and other derivatized applications, such as ChIP-chip (DeRisi *et al.*, 1997; Morley *et al.*, 2004; Pease *et al.*, 1994; Schadt *et al.*, 2003; Schena *et al.*, 1995). However, RNA expression levels do not always correlate with protein expression levels, and biological functions are carried out primarily by proteins rather than nucleic acids (Gygi *et al.*, 1999; Lueking *et al.*, 2005b). Therefore, it was the next logical step to develop a miniaturized protein-centered device, namely protein microarrays, for studies of protein functionalities in a high-throughput and highly flexible fashion.

A protein microarray, also known as a protein chip, is formed by immobilization of thousands of different proteins (e.g., antigens, antibodies, enzymes, substrates, etc.) in discrete spatial locations at a high-density solid surface (typically glass) (Smith *et al.*, 2005; Tao *et al.*, 2007). On the basis of their applications, protein microarrays can be classified into two types: analytical and functional protein microarrays (Fig. 4.1). Analytical protein microarrays are usually composed of well-characterized biomolecules with specific binding activities, such as antibodies, to analyze the components of complex biological samples (e.g., serum and cell lysates) or to determine whether a sample contains a specific protein of interest. They have been used for protein expression profiling, biomarker identification, cell surface marker/glycosylation profiling, clinical diagnosis, and environmental/food safety analysis (Kumble, 2003). On the other hand, functional protein microarrays are constructed by printing a large number of individually purified proteins and are mainly used to comprehensively query biochemistry properties and activities of those immobilized proteins. In principle, it is feasible to print arrays composed of virtually all annotated proteins of a given organism, effectively comprising a whole-proteome microarray. Functional protein microarrays have been successfully applied to identify protein–protein, protein–lipid, protein–antibody, protein–small molecules, protein–DNA, protein–RNA, lectin–glycan, and lectin–cell interactions, and to identify substrates or enzymes in phosphorylation, ubiquitylation, acetylation, and nitrosylation, as well as to profile immune response. In this chapter, we will mainly focus on the fabrication and application of functional protein microarrays in basic and clinical research.

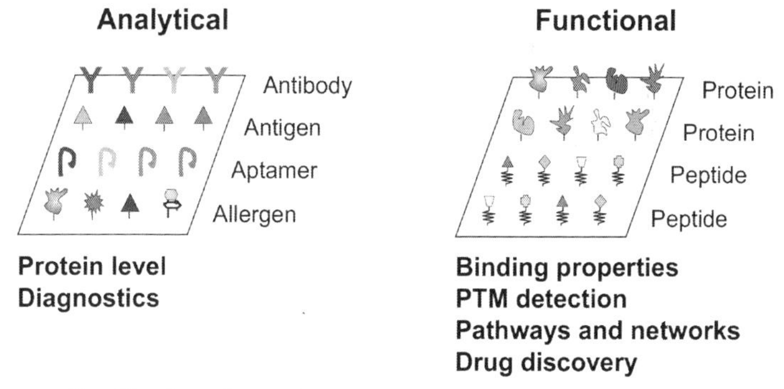

Figure 4.1 ***Classification of protein microarrays.*** Protein microarrays are of two types: analytical and functional protein microarrays. Left: Analytical protein microarrays are constructed using biomolecule with specific binding property, such as antibodies, antigens, and aptamers. Right: Functional protein microarrays are formed by immobilization of individually purified proteins or synthetic peptides. The major applications of both types are listed below. For color version of this figure, the reader is referred to the online version of this book.

II. FABRICATION OF FUNCTIONAL PROTEIN MICROARRAYS

Because the biochemical properties of DNA molecules are essentially the same, the same chemistries can be applied to either immobilize to or synthesize *in situ* DNA strands on a solid surface (DeRisi *et al.*, 1997; Pease *et al.*, 1994). Therefore, the design and construction of oligonucleotide DNA microarrays are relatively straightforward. However, the protein world is manifested by much more complicated biochemistries, which are reflected by vast differences in protein size, shape/conformation, charge, stability, and hydrophobicity, to name a few. Furthermore, many proteins are known to require proper partners to be able to execute their biochemical activities. This implies that the fabrication and analysis of protein microarrays is substantially more challenging than that of DNA microarrays. Unlike DNA or RNA molecules, full-length proteins cannot be directly synthesized *in vitro* at high efficiency. Although *in vitro* synthesis of peptides has been feasible for decades, it still suffers from low yield, high cost, and effective limitation to short sequences. Moreover, the vast majority of proteins must be correctly folded and modified to be functional during and after translation, which may require a complex molecular machinery of chaperones and other accessory molecules that cannot be fully recapitulated *in vitro*. Therefore, the development of a high-throughput method that allows for purifying proteins under native conditions is the key to fabricate functional protein microarrays of high content.

A. High-Throughput Protein Production

Although many methods have been developed to purify proteins from both eukaryotic and prokaryotic systems, the main hurdle has been the difficulty in producing a large number of different proteins needed for construction of a truly high-content, functional protein microarray. Obviously, a readily useable high-throughput protocol for parallel production of thousands of different proteins is the key to this challenge.

An early attempt led by the Lehrach group was to express human proteins in *Escherichia coli* using a library consisting of random cDNAs (Bussow *et al.*, 1998). Individual cDNA clones of this library were robotically arrayed onto polyvinylidene difluoride (PVDF) membrane laid on top of agar media and allowed to grow to full size. These cells were then lysed *in situ* to extract proteins. The usefulness of such an array was first demonstrated by incubation with a labeled test protein to identify interacting partners (Bussow *et al.*, 1998). Strictly speaking, besides the fact that only one-sixth of the cDNAs are in the proper reading frames, those correctly expressed human proteins bound to the nitrocellulose membranes were not purified—the majority of the proteins in every spot were bacterial proteins. Furthermore, the proteins were neither unique nor in their native conformation, given the redundancy of the library and denaturing conditions used to break the bacteria open. Though powerful as a screening technique in early days, this particular experimental strategy had limited general applications (Holt *et al.*, 2000; Lueking *et al.*, 1999).

To overcome these hurdles, Zhu *et al.* (2001) in the Snyder group created a high-throughput protein purification protocol in the budding yeast (Fig. 4.2). Using a homologous recombination-based strategy, more than 5800 full-length yeast open reading frames (ORFs) were cloned into

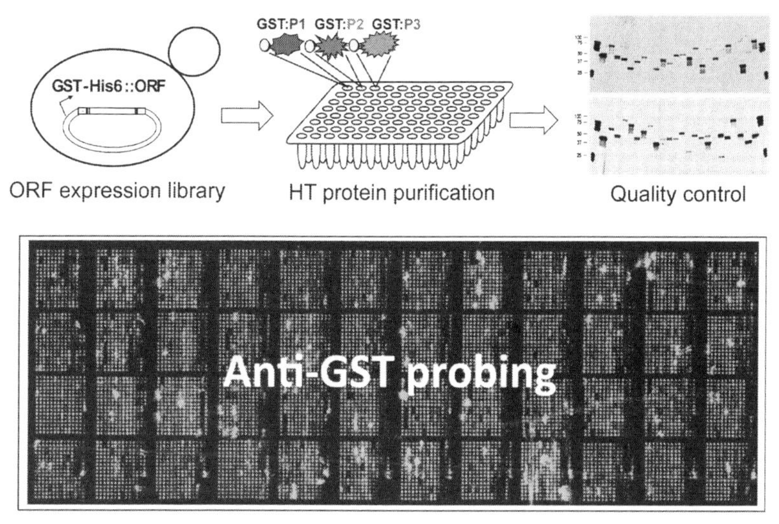

Figure 4.2 ***Fabrication of high-content functional protein microarrays.*** Four major steps are involved to construct a functional protein microarray of high content. First, a high-quality ORF expression library is constructed to allow inducible overexpression of GST-His6 fused recombinant proteins in yeast. Second, a high-throughput protein purification protocol is applied to individually purify thousands of proteins from yeast. The purified proteins are stored in 384-well format. Third, silver stain and immunoblot analysis are employed to evaluate the quality and quantity of the purified proteins. Finally, when purified proteins pass the quality control, they are spotted in duplicate to glass slides using a robot microarrayer. The quality of printing is then tested by anti-GST probing. An image of a human protein microarray probed with anti-GST is shown in the lower panel. For color version of this figure, the reader is referred to the online version of this book.

a yeast expression vector that, upon galactose induction, produces glutathione S-transferase (GST)-tagged N-terminal fusion proteins. The purification protocol took advantage of both a 96-well format and immobilized affinity chromatography. This strategy allowed parallel purification of unprecedented numbers of proteins—up to 1152 per day. The success of this approach is built upon several unique aspects. First, it utilizes a eukaryotic expression system that both generates high levels of recombinant proteins and tends to produce a high fraction of soluble proteins. Compared with bacterial expression systems, in which a large fraction of recombinant proteins end up in inclusion bodies, this is a huge advantage when a large number of eukaryotic proteins are being generated. Second, the expression of recombinant proteins is only induced over about two total cell cycles, which greatly reduces toxicity and cell death. Third, a foreign eukaryotic protein purified from yeast is more likely to be active because post-translational modifications necessary for function are more likely to occur correctly than in either bacteria or a cell-free system. Forth, the use of an N-terminal GST tag helps protein fold correctly and therefore improve its stability and solubility. Other commonly used tags include the so-called TAP tag, MPB, and Hisx6, to name a few. In fact, the same group later went on to build a TAP-tagged yeast ORF collection and purified >5,000 yeast proteins (Gelperin *et al.*, 2005).

Another commonly used expression system is *E. coli*. The procedures for automatic high-throughput protein expression/purification using the 6xHis tag have been developed by the Zhu group (Chen *et al.*, 2008). Once the bacterial culture is prepared, all 4000 proteins can be purified within a single day. The subsequent protein purification takes advantage of immobilized Ni-NTA affinity chromatography (Hochuli *et al.*, 1987). The 6xHis tag usually does not alter the properties of the fusion proteins, and the increment of molecular weight is less than 1 kDa. Furthermore, it is selective and stable even under severe denaturing conditions (Joshi *et al.*, 2000; Mukhija *et al.*, 1995).

Despite the fact that high-throughput protein production in both prokaryotes and eukaryotes is now increasingly feasible, these protocols are labor intensive and costly. Aside from the cost of protein production, fabrication of a proteome microarray requires construction of an expressible collection of full-length ORFs, which can be both challenging and expensive when dealing with higher eukaryotes with a large number of genes, such as humans. To explore alternative approaches, several groups have attempted to test the *in vitro* transcription/translation systems, such as the *E. coli*, wheat germ, and rabbit reticulocyte systems. In these systems, proteins can be expressed directly from cDNA templates (Allen and Miller, 1999), which can be obtained through polymerase chain reaction amplification without the lengthy and costly process of subcloning. For example, the *E. coli* cell-free protein expression system has been used to synthesize proteins in a 96-well format (Murthy *et al.*, 2004), and the improved wheat germ cell-free protein synthesis system has been applied to the *in vitro* expression of 13,364 human proteins (Goshima *et al.*, 2008). More recently, the Felgner group has published a series of articles describing fabrication of protein microarrays in a variety of bacteria by directly spotting *in vitro* translated protein mixtures to glass (Crompton *et al.*, 2010; Liang *et al.*, 2011). However, although these systems can significantly decrease the reaction volume required for generation of recombinant proteins (Angenendt *et al.*, 2004), the impurity of the translated proteins limits their applications.

Such systems can also be applied to directly synthesize proteins on glass slides to fabricate so-called "*in situ* protein microarrays." In the Protein *In Situ* Array (PISA) method, proteins are expressed directly from DNA *in vitro* and become attached to the array surfaces through recognition of a sequence that serves as an affinity tag (He and Taussig, 2001). Similarly, in the Nucleic Acid Programmable Protein Array (NAPPA) technology, biotinylated cDNA plasmids encoding proteins as GST fusions are printed onto avidin-coated slides, together with anti-GST antibodies as the capture molecules (Ramachandran *et al.*, 2004). The cDNA array is then incubated with rabbit reticulocyte lysate to express the proteins, which become trapped by the antibodies adjacent to each DNA spot. Recently, NAPPA has been successfully expanded to high-density arrays of 1000 different proteins (Ramachandran *et al.*, 2008). In addition, Tao and Zhu (2006) developed a different method in which ribosomes are installed at the end of an RNA template to allow for the capture of the nascent polypeptides by a puromycin moiety that is grafted at one end of an oligonucleotide immobilized on a solid surface.

Another similar method is called DNA Array to Protein Array (DAPA), in which proteins are synthesized between two glass slides: one of which is arrayed with DNA, whereas the other carries a specific affinity reagent to capture the proteins (He *et al.*, 2008). In this approach, tagged proteins are synthesized in parallel from the DNA array, spread across the gap between the two slides, and then bound to the tag-capturing reagents on the other slide to form a protein array. Unlike the NAPPA method in which proteins are present together with DNA and the DNA array can only be used once, DAPA generates multiple copies of "pure" protein arrays on a separate surface from the same DNA template, with at least 20 copies capable of being produced from a single template.

Because proteins must fold correctly in order to be active and because proteins are prone to inactivation due to loss of their native conformations (e.g., exposure to denaturing conditions during purification), it is better to express proteins of interest in cells and purify them under native conditions.

B. Surface Chemistry

Choosing a proper surface for protein immobilization is crucial to the success of any assay performed using protein microarrays. An ideal surface should be able to retain protein functionality with relatively high signal-to-noise ratios and possess both high protein-binding capacity and long shelf life (Smith *et al.*, 2005; Tao *et al.*, 2007). Glass slides covered with PVDF, nitrocellulose membrane, or polystyrene were popular for protein microarray fabrication in the early days of the technology (Bussow *et al.*, 1998; Holt *et al.*, 2000; Lueking *et al.*, 1999). However, PVDF and polystyrene are relatively soft, allowing lateral spread of printed proteins, and hence limited density of proteins to be printed. Nitrocellulose membranes, in addition, tend to generate high background and low signal-to-noise ratio for most applications.

To bypass these shortcomings, researchers developed three-dimensional matrix arrays, in which glass slides are coated with polyacrylamide or agarose to form a porous hydrophilic matrix in which proteins or antibodies are trapped within the pores and lateral diffusion is restricted, reducing the size of printed protein spots and thus increasing the maximal complexity of the array (Afanassiev *et al.*, 2000; Guschin *et al.*, 1997). Protein activity is generally well preserved in such matrix arrays, and their protein binding capacity is relatively high. For instance, Zhu *et al.* (2000) utilized soft lithography to generate nanowells on a polydimethylsiloxane sheet placed on top of microscope slides. These nanowell chips were used to immobilize substrate proteins to profile phosphorylation specificity of 119 kinases encoded by budding yeast. The open structure of nanowells provides physical barriers and allows for sequential adding of different buffers, which is critical for multistep experiments. The main disadvantage of this method is the requirement of specialized equipment needed to load nanowells at high density.

Other researchers printed proteins, antigens, or antibodies directly onto plain glass slides, which are usually coated with a bifunctional cross-linker with two functional groups, one reacting with the glass surface and the other with the desired proteins. For example, Schweitzer *et al.* (2000) demonstrated in their study that protein microarrays fabricated on glass surface possess high sensitivities, wide dynamic range, and decent spot-to-spot reproducibility. MacBeath and Schreiber (2000) demonstrated with three proteins that thousands of protein spots could be immobilized to aldehyde-activated plain glass surfaces to form a high-density protein microarray that was suitable for a range of different classes of assays.

C. Protein Immobilization

The physical and chemical properties of different proteins vary greatly, and protein activities are closely related to their structures. Therefore, the development of a stable universal immobilization method that does not change protein structures is one of the difficulties of protein microarray fabrication. To this end, several different methods have been used for protein immobilization on solid carrier surfaces, such as noncovalent adsorption, covalent binding, and affinity capture.

Noncovalent adsorption provides both high protein capacities and low impact on protein structures but cannot control the amount and orientation of immobilized proteins. Thus, the reaction efficiency, accuracy, and reproducibility of arrays produced in this manner are variable. Covalent binding, on the other hand, results in chemically cross-linked proteins via reactive residues (e.g., lysine and cystine) to surface-grafted ligands, such as aldehyde, epoxy, reactive ester, etc. (MacBeath and Schreiber, 2000; Templin *et al.*, 2002; Ziauddin and Sabatini, 2001). Lee *et al.* (2003) developed novel calix crown derivatives as a ProLinker that permits efficient immobilization of captured proteins on solid matrixes, and the immobilized proteins showed both consistent directionality and functionality. Covalent binding is suitable for immobilization of a wide range of proteins with strong conjunctions to the carrier surfaces. However, the modification of chemical groups can sometimes both alter the activities of target proteins and their binding to specific ligands.

Affinity capture is an attractive way to immobilize proteins that avoid many of the shortcomings of the previously detailed approaches. For example, biotinylated proteins have been used for protein immobilization to streptavidin-coated slides. The use of genetically encoded affinity tags, which can be fused to target proteins and bind to a specific slide surface, is an analogous approach. For example, 6xHis-tags have been utilized to immobilize proteins on nickel-NTA-coated glass slides (Zhu *et al.*, 2001). Presumably, affinity-based protein immobilization should result in immobilization of proteins in relatively uniform orientation with minimum interruption of protein structure and thus, may be the best approach to for preserving the structure and function of printed proteins. One important caveat to bear in mind, however, is that the incorporation of affinity tags may alter the protein structures.

One way to deal with this challenge was demonstrated by Zhang *et al.* (2005), who developed a flexible polypeptide scaffold consisting of a surface immobilization domain and a protein capture domain, which allows much greater flexibility in the immobilization of proteins on a microarray. Wacker *et al.* (2004) compared the DNA-directed immobilization (DDI) method with both direct spotting and with biotin–streptavidin affinity immobilization for antibodies. DDI is based on the self-assembly of semisynthetic DNA–streptavidin conjugates that convert a DNA oligomer array into an antibody array (Niemeyer *et al.*, 1994). DDI and direct spotting showed the highest fluorescence intensities. DDI also performed the best in spot homogeneity and intra- and interexperimental reproducibility. Moreover, DDI required the lowest amount of antibodies, at least 100-fold less than direct spotting. The drawback of DDI is that proteins have to be linked to DNA prior to immobilization, which increases the workload involved in generating microarrays.

The orientation of immobilized proteins may influence both their activity and their affinity for specific ligands. Peluso *et al.* (2003) compared randomly versus specifically oriented capture agents based on both full-sized antibodies and Fab' fragments. The specific orientation of capture agents consistently increased the analyte-binding capacity of the surfaces up to 10-fold relative to surfaces with randomly oriented capture agents. When specifically oriented, Fab' fragments formed a dense monolayer and 90% of them were active, whereas randomly attached Fab's both packed at lower density and had lower specific activity.

III. SIGNAL DETECTION

In addition to optimized surface modification and optimized reaction condition, the detection sensitivity of samples bound on microarrays is another key parameter in the design of protein microarray assays. There are two basic detection methods: label-dependent and label-free detections.

A. Label-Dependent Detection Methods

Radioisotopes and fluorescent dyes are the two most common labeling methods for signal detection in protein microarray assays. Fluorescent dyes, such as Cy-3/5 and their equivalent, have been used as a popular labeling method. Because most good dyes have relatively narrow excitation and emission spectra, multicolor scheme can be readily implemented for simultaneous detection and direct comparison of different samples, both reducing cost and avoiding chip-to-chip variation. Semiconductor quantum dot labeling, which is brighter and more stable than organic dyes, has also been applied to protein microarrays (Shingyoji *et al.*, 2005; Zajac *et al.*, 2007).

In addition to fluorescent labeling, Huang (2001) detected multiple cytokines on an antibody array with enhanced chemiluminescence, providing an alternative detection method. Enzymatic signal amplification is also a valuable labeling method. Rolling circle amplification (RCA) has been developed for protein microarray assays. For low abundance protein samples, the sensitivity of traditional fluorescence or chemiluminescence detection is relatively low, whereas RCA can detect captured proteins at femtomole level and is promising to improve the sensitivity of fluorescent detection (Lizardi *et al.*, 1998; Schweitzer *et al.*, 2000; Schweitzer *et al.*, 2002; Shao *et al.*, 2003; Zhou *et al.*, 2004). Tyramide signal amplification is another way to amplify signals with enzymes, which utilizes the horseradish peroxidase conjugated on secondary antibodies to convert the labeled substrates (tyramide) into short-lived extremely reactive intermediates, which then very rapidly react with and covalently bind to adjacent proteins (Varnum *et al.*, 2004).

For certain types of biochemical assays, especially enzymatic reactions, use of radioisotopes is the only detection method available (see below for more details). They still offer the most sensitive and reliable detection of posttranslation modification events when there is a lack of high-quality and high-affinity detection reagents, such as antibodies. We and others have successfully applied ^{32}P-, ^{33}P-, and ^{14}C-labeled substrates to detect protein phosphorylation and acetylation events (Lin *et al.*, 2009; Lu *et al.*, 2011; Ptacek *et al.*, 2005; Zhu *et al.*, 2009).

B. Label-Free Detection Methods

One obvious disadvantage of label-dependent detection is the requirement of either manipulating structure of a probe or a specific antibody. It is not amenable to real-time label-free detection, which can provide important information when analyzing reaction dynamics. Therefore, label-free detection methods have also been investigated for protein microarrays. Optical techniques of various types are emerging as an important tool for mentoring the dynamics of biomolecule interactions on a solid surface. For instance, Imaging Surface Plasmon Resonance Spectroscopy (SPR) (Nelson *et al.*, 1999; Thiel *et al.*, 1997), Imaging Optical Ellipsometry (OE) (Wang and Jin, 2003), and Reflectometric Interference Spectroscopy (Piehler *et al.*, 1997) are three label-free optical techniques that in essence measure the same optical dielectric response of a thin film and therefore detect changes of physical or chemical properties of the thin film, such as thickness and mass density during biochemical reactions.

As compared with the above three methods, the oblique-incidence reflectivity difference (OIRD) technique is a more sensitive form of ellipsometry that measures the difference in reflectivity between *s*- and *p*-polarized light (Chen *et al.*, 2001; Landry *et al.*, 2004). Recently, the OIRD technique has been applied to detect DNA hybridization and protein–protein interactions in a microarray format in a real-time fashion, and these studies demonstrated its potential as a high-throughput detection method that can obtain association and dissociation rates of biomolecule interactions (Lu *et al.*, 2010; Wang *et al.*, 2010). This is an extremely sensitive detection method: it has a time resolution of 20 μs, a space resolution (i.e., thickness) of 0.4 nm, and a detection limit of 14 fg of protein per spot. In addition, it also shares other advantages of the SPR and OE methods, such as noncontacting damage- and label-free detection (Fei *et al.*, 2008; Lu *et al.*, 2010).

The principle of OIRD-based detection is illustrated in Fig. 4.3. First, a *p*-polarized He-Ne laser beam ($\lambda = 632.8$ nM) passes through a photoelastic modulator to induce oscillation between *p*- and *s*-polarization at a frequency of 50 kHz. Second, after passing through a phase shifter, the resultant beam is incident on the microarray surface at an oblique angle theta (θ_{inc}). Finally, the first $I(\Omega)$ and second harmonics $I(2\Omega)$ of the reflected beam intensity are simultaneously monitored by two digital lock-in amplifiers (Lu *et al.*, 2010). The difference caused by changes in reflectivity between the *s*- and *p*-polarized light, namely the OIRD signal, is $\Delta p - \Delta s$, composed of both real and imaginary components. Because the imaginary component, which is proportional to the first $I(\Omega)$, is more sensitive, the OIRD signal is determined as "$\mathrm{Im}\{\Delta p - \Delta s\}$" (Formula I), which is dependent on the incident angle (θ_{inc}) and the dielectric constants of the ambient, protein, and substrate of the microarray (i.e., glass) (Wen *et al.*, 2010).

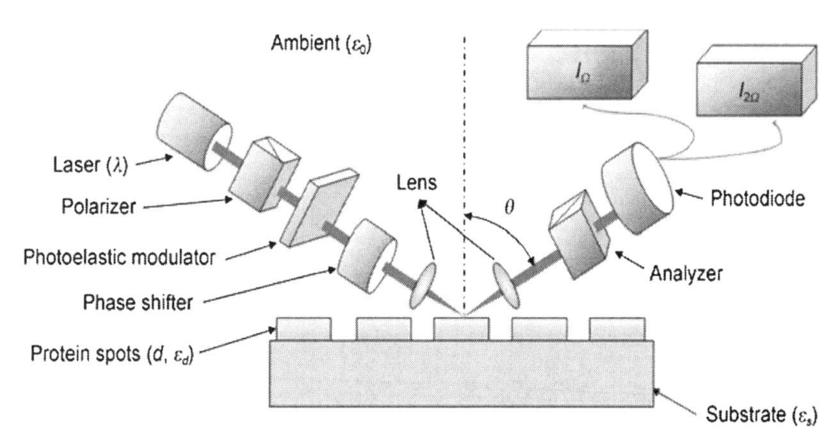

Figure 4.3 ***Principle of the OIRD method.*** First, a *p*-polarized He-Ne laser beam ($\lambda = 632.8$ nM) passes through a photoelastic modulator to induce oscillation between *p*- and *s*-polarization at a frequency of 50 kHz. Second, after passing through a phase shifter, the resultant beam is incident on the microarray surface at an oblique angle theta (θ_{inc}). Finally, the first $I(\Omega)$ and second harmonics $I(2\Omega)$ of the reflected beam intensity are simultaneously monitored by two digital lock-in amplifiers.

Formula I

$$\mathrm{Im}\{\Delta_p - \Delta_s\} = -\beta \frac{(\varepsilon_d - \varepsilon_s)(\varepsilon_d - \varepsilon_0)}{\varepsilon_d} d,$$

$$\beta = \frac{4\pi\varepsilon_s(\tan\theta_{\mathrm{inc}})^2\cos\theta_{\mathrm{inc}}}{\varepsilon_0^{1/2}(\varepsilon_s - \varepsilon_0)(\varepsilon_s/\varepsilon_0 - (\tan\theta_{\mathrm{inc}})^2)\lambda}, \quad (1)$$

The OIRD system is attractive in several ways. First, it was first designed in the format of microarray and, therefore, readily to be applied to protein microarray assays. Second, it is promising to be developed as an extremely high-throughput method as it has already been able to detect approximately 10,000 protein spots at once (Lv et al., personal communication). Third, unlike the SPR system, it is not restricted to a particular surface type, which makes it much more flexible for various types of biochemical assays.

Finally, mass spectrometry has also been used for detecting ligands bound to individual proteins printed on protein microarrays, with such approaches as MALDI-MS, SELDI-TOF-MS, and MALDI-TOF-MS used for this purpose (Diamond *et al.*, 2003; Evans-Nguyen *et al.*, 2008; Gavin *et al.*, 2005). The analysis is rapid and simple, requires small sample amount, and can be used for direct detection of analytes bound from complex samples, such as urine, serum, plasma, and cell lysates. Atomic force microscopy (AFM) uses surface topological changes to identify the analytes bound on the array (Lee *et al.*, 2002; Yan *et al.*, 2003). More specifically, AFM detects the increase in height of the proteins/antibodies on the array and thus is able to measure binding interactions.

IV. APPLICATIONS OF FUNCTIONAL PROTEIN MICROARRAYS

A. Development of new Assays

Unlike the DNA/oligo microarray or analytical protein microarrays, functional protein microarrays provide a flexible platform that allows development and detection of a wide range of protein biochemical properties. To date, well-developed assays include detection of various types of protein–ligand interactions, such as protein–protein, protein–DNA, protein–RNA, protein–lipid, protein–drug, and protein–glycan interactions (Chen *et al.*, 2008; Hall *et al.*, 2004; Ho *et al.*, 2006; Hu *et al.*, 2009; Huang *et al.*, 2004; Kung *et al.*, 2009; MacBeath and Schreiber, 2000; Popescu *et al.*, 2007; Zhu *et al.*, 2001; Zhu *et al.*, 2007), and identification of substrates of various classes of enzymes, such as protein kinase, ubiquitin E3 ligase, and acetyltransferase, to name a few(Lin *et al.*, 2009; Lu *et al.*, 2008; Ptacek *et al.*, 2005; Schnack *et al.*, 2008; Zhu *et al.*, 2000).

During the development of various assay types, it became obvious that surface chemistry plays an important role in the success of a new assay (Table 4.1). For example, protein–DNA interactions were first performed on the yeast proteome microarrays on a nitrocellulose surface (i.e., FAST slide) with randomly shared yeast genomic DNA fragments that were labeled with Cy5 (Hall *et al.*, 2004). Later, Hu *et al.* (2009) also found that the FAST slide, among other tested surfaces, produced the best signal-to-noise ratio for DNA-binding assays. In another example, when our group was developing protein acetylation reactions using ^{14}C-labeled Ac-CoA as a donor, we first tested the NuA4 acetylation reaction using histone H3 and H4 as substrates on FAST slides, as well as aldehyde- and Ni-NTA-coated slides (Lin *et al.*, 2009; Lu *et al.*, 2011). The results were clear that both FAST and nickel surfaces worked, but FAST surface produced better signal-to-noise ratios (Fig. 4.4). However, FAST surface was not suitable for phosphorylation reactions because the background noises were too high (Ptacek *et al.*, 2005;

Table 4.1 Effects of surface chemistry to protein microarray assays

Assay \ Surface	Fullmoon	Aldehyde	FAST	PATH	Schott	Ni-NTA	Detection method	Ref.
Protein-protein	√	√	√			√	antibody	15,28,29
Protein-lipid			√				Fluorescence	15
Protein-DNA			√				Fluorescence	59-61
Protein-RNA			√				Fluorescence	75
Protein-drug	√						Biotin	76
Lectin-live cell					√		Fluorescence	80-82
Glycan-protein	√		√				Fluorescence	77
Phosphorylation	√	√		√			^{32}P, ^{33}P	52,55
Acetylation			√			√	^{14}C	53,54
Ubiquitylation	√						antibody	79
SUMOylation	√						antibody	103
Nitrosylation	√						biotin	87
Serum profiling	√		√	√			antibody	85-89

Zhu *et al.*, 2009). Another interest case is to develop an assay for profiling cell surface glycans on a lectin microarray. We and others found that so far the only proper surface for this type of binding assays is a commercial Schott slide, although the exact surface chemistry is not revealed (Hsu *et al.*, 2006; Pilobello *et al.*, 2007; Tao *et al.*, 2008). Several reasons may be accounted for the importance of surface chemistry. First, for low-affinity binding assays (e.g., protein–DNA interactions), a porous surface (e.g., FAST) is likely to retain more proteins and hence improving sensitivity. Second, when radioisotope-labeled small molecules are used, it is important to completely remove unincorporated radioisotopes from the surface to reduce background noise. This might explain why phosphorylation does not work well on FAST surface. Third, in the case of using live cells to probe a lectin microarray, the grafted chemical ligands must not be too repulsive to cells. Other factors, such as protein conformation and stability, can also be affected by surface chemistry. Therefore, whenever a novel assay is to be developed, a variety of surfaces should be tested first in a pilot study.

Application of these assays has had a profound impact on a wide range of research areas. This is especially true when they are used in large-scale high-throughput projects, exemplified in both network construction and biomarker identification (see below and Table 4.2).

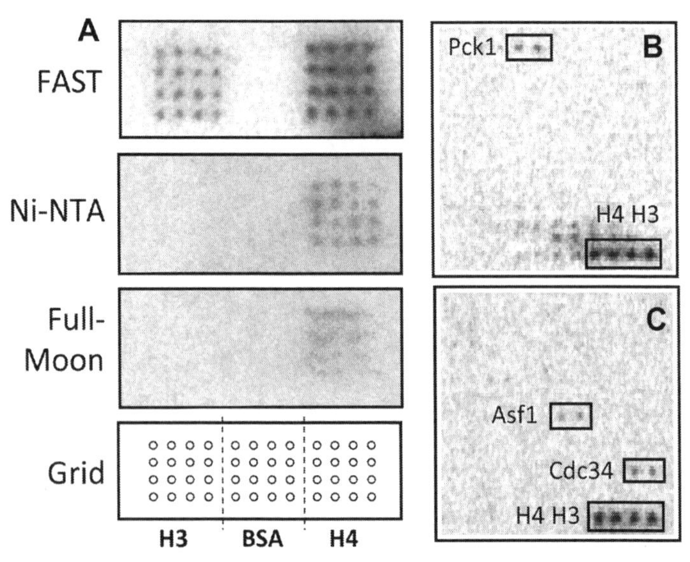

Figure 4.4 ***Effects of surface chemistry on assay development.*** (A) A pilot experiment to optimize the reaction conditions for protein acetylation in a microarray format. In each reaction, the yeast NuA4 acetyltransferase complex was added to an acetylation reaction mixture containing ^{14}C-Ac-CoA and incubated with histone H3 and H4 spotted on three different surfaces. Bovine serum albumin (BSA) was also included as a negative control. The acetylation signals were detected by long exposure to X-ray film. (B, C) Examples of newly identified non-histone substrates. For color version of this figure, the reader is referred to the online version of this book.

B. Detection of Protein-Binding Properties

1. Protein–protein interaction

Among the first applications of protein microarrays was in the analysis of protein–protein and protein–lipid interactions, where test ligands were directly or indirectly labeled with fluorescent dyes. For example, Zhu *et al.* (2001) developed the first proteome microarray composed of approximately 5,800 recombinant yeast proteins (>85% of the yeast proteome) and identified binding partners of calmodulin and phosphatidylinositides (PIPs). They first incubated the microarrays with biotinylated bovine calmodulin and discovered 39 new calmodulin binding partners. In addition, using liposomes as a carrier for various PIPs, they identified more than 150 binding proteins, more than 50% of which were known membrane-associated proteins. Popescu *et al.* (2007) developed a protein microarray containing 1,133 *Arabidopsis thaliana* proteins and also used it to globally identify proteins bind to calmodulins or calmodulin-like proteins in *Arabidopsis*. A large number of previously known and novel targets were identified, including transcription

Table 4.2 Application of functional protein microarrays in large-scale projects

Type of assay	Type of array	Type of probe	No. of probe	Application	Reference
Protein–peptide interaction	Human SH2 and PTB domain array	Peptide	61	Protein interaction network	(Jones *et al.*, 2006)
Protein–DNA interaction	Yeast TF array	DNA motif	75	Protein–DNA interaction Network	(Ho *et al.*, 2006)
	Human TF array	DNA motif	460	Protein–DNA interaction Network	(Hu *et al.*, 2009)
Kinase assay	Yeast proteome array	Protein kinase	87	Phosphorylation network	(Ptacek *et al.*, 2005)
Antigen–antibody interaction	Coronavirus array	SARS patient sera	602	Biomarker identification	(Hu *et al.*, 2007)
	E. coli proteome array	Inflammatory bowel disease (IBD) patient sera	134	Biomarker identification	(Xie *et al.*, 2010)
	Human protein array	Ovarian cancer patient sera	60	Biomarker identification	(Jones *et al.*, 2006)
	Human protein array	Alopecia areata patient sera	44	Biomarker identification	(Foster *et al.*, 2009)
	Human protein array	Autoimmune hepatitis (AIH) patient sera	278	Biomarker identification	(Robinson *et al.*, 2002)

factors (TFs), receptor and intracellular protein kinases, F-box proteins, RNA-binding proteins, and proteins of unknown function. Alternative approaches to identifying protein–protein interactions, such as the yeast two-hybrid system and protein complex purification coupled with mass spectrometry analysis, are well-established, however, and are used as standard high-throughput methods to detect protein–protein interactions in higher eukaryotes (Krogan *et al.*, 2006; Vidal *et al.*, 1996). Thus, while protein microarray-based approaches provide a rapid approach to characterizing protein–protein interactions, they have much competition in this arena.

2. Protein–peptide interaction

MacBeath *et al.* fabricated protein domain microarrays to investigate protein–peptide interactions in a semiquantitative fashion that might play an important role in signaling (Jones *et al.*, 2006). They constructed an array by printing 159 human Src homology 2 (SH2) and phosphotyrosine binding (PTB) domains on the aldehyde-modified glass substrates and incubated the arrays with 61 peptides representing tyrosine phosphorylation sites on the four ErbB receptors. Eight concentrations of each peptide (10–5 mM) were tested in the assay, allowing quantitative measurement of the binding affinity of each peptide to its protein ligand.

3. Protein–DNA interaction

Protein microarrays have also been applied extensively and productively to characterize protein–DNA interactions (PDIs). In an earlier study, Snyder *et al.* screened for novel DNA-binding proteins by probing the yeast proteome microarrays with fluorescent labeled yeast genomic DNA (Hall *et al.*, 2004). Of the approximately 200 positive proteins, half were not previously known to bind to DNA. By focusing on a single yeast gene, *ARG5,6*, encoding two enzymes involved in arginine biosynthesis, they discovered that it bound to a specific DNA motif and associated with specific nuclear and mitochondrial loci *in vivo*.

In a later report, the Snyder and Johnston groups constructed a protein microarray with 282 known and predicted yeast TFs to identify their interactions with 75 evolutionarily conserved DNA motifs (Ho *et al.*, 2006). Over 200 specific PDIs were identified, and more than 60% of them are previously unknown. The binding site of a previously uncharacterized DNA-binding protein, Yjl103p, was defined, and a number of its target genes were identified, many of which are involved in stress response and oxidative phosphorylation.

Our team developed a bacterial proteome microarray composed of 4,256 proteins encoded by the *E. coli* K12 strain (approximately 99% coverage of the proteome) using a bacterial high-throughput protein purification protocol (Chen *et al.*, 2008). To demonstrate the usefulness, end-labeled double-stranded DNA probes carrying a basic or mismatched base pairs were used to identify proteins involved in DNA damage recognition. A small number of proteins were specifically recognized by each type of the probes with high affinity. Two of them, YbaZ and YbcN, were further characterized to encode base-flipping activity using biochemical assays.

Recently, our group also undertook a large-scale analysis of human PDIs using a protein microarray composed of 4,191 unique human proteins in full length, including approximately 90% of the annotated TFs and a wide range of other protein categories, such as RNA-binding proteins, chromatin-associated proteins, nucleotide-binding proteins, transcription coregulators, mitochondrial proteins, and protein kinases (Hu *et al.*, 2009). The protein microarrays were probed with 400 predicted and 60 known DNA motifs, and a total of 17,718 PDIs were identified. Many known PDIs and a large number of new PDIs for both well-characterized and predicted TFs were recovered, and new consensus sites for over 200 TFs were determined, which doubled the number of previously reported consensus sites for human TFs (Hu *et al.*, 2009; Xie *et al.*, 2010). Surprisingly, over 300 proteins that were previously unknown to specifically interact with DNA showed sequence-specific PDIs, suggesting that many human proteins may bind specific DNA sequences as a moonlighting function. To further investigate whether the DNA-binding activities of these unconventional DNA binding proteins (uDBPs) were physiologically relevant, we carried out in-depth analysis on a well-studied protein kinase, Erk2, to determine the potential mechanism behind its DNA-binding activity. Using a series of *in vitro* and *in vivo* approaches, such as electrophoretic mobility shift assay (EMSA), luciferase assay, mutagenesis, and chromatin immunoprecipitation (ChIP), we demonstrated that the DNA-binding activity of Erk2 is independent of its protein kinase activity and it acts as a transcription repressor of transcripts induced by interferon gamma signaling (Hu *et al.*, 2009). Other than Erk2, many other uDBPs show sequence-specific DNA-binding activity, and more intriguingly, many of their consensus sequences are highly similar to those recognized by annotated TFs (Fig. 4.5). This observation suggests that these uDBPs may synergistically work with the TFs to achieve highly accurate transcription regulation.

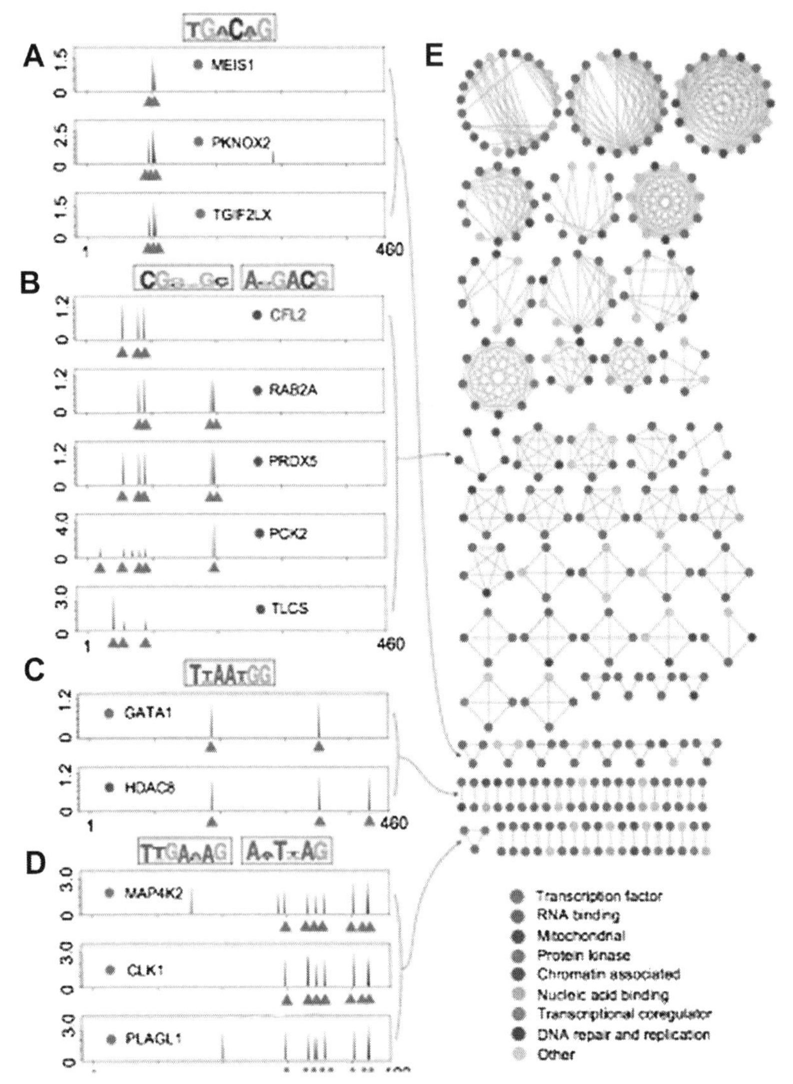

Figure 4.5 ***Similar consensus sites are recognized by both TFs and uDBPs.*** (A–D) Examples of proteins sharing similar DNA binding profiles. Each peak represents normalized signal intensity of a specific DNA motif probe, with individual motifs organized along the X-axis by sequence similarity. Binding peaks used to generate the major logos (outlined in red) are indicated by red triangles. For proteins that recognize more than one logo (outlined in blue), binding peaks for the second logo are indicated in blue. (E) Correlation network for proteins with highly similar DNA binding profiles (see Supplementary Methods for construction of the network). Protein class is indicated by colored dots. For interpretation of the references to color in this figure legend, the reader is referred to the online version of this book.

4. Protein–Small molecule interaction

Discovering new drug molecules and drug targets is another field in which protein microarrays have shown its potential. For example, Huang *et al.* (2004) incubated biotinylated small-molecule inhibitors of rapamycin (SMIRs) on the yeast proteome microarrays and obtained the binding profiles of the SMIRs across the entire yeast proteome. They identified candidate target proteins of the SMIRs, including Tep1p, a homologue of the mammalian PTEN tumor suppressor, and Ybr077cp (Nir1p), a protein of previously unknown function, both of which are validated to associate with PI(3,4)P2, suggesting a novel mechanism by which phosphatidylinositides might modulate the target of rapamycin (TOR) pathway.

5. Protein–RNA interaction

The yeast proteome microarray has been used to identify specific RNA-binding proteins for antiviral activities (Zhu *et al.*, 2007). In these experiments, arrays were incubated with a fluorescently tagged small RNA hairpin containing a clamped adenine motif, which is required for the replication of Brome Mosaic Virus (BMV), a plant-infecting RNA virus that can also replicate in the budding yeast. Two of the candidate proteins, Pseudouridine Synthase 4 (Pus4) and the Actin Patch Protein 1 (App1), were further characterized in *Nicotiana benthamiana*. Both of them modestly reduced BMV genomic plus-strand RNA accumulation and dramatically inhibited the spread of BMV in plants.

6. Protein–Glycan interaction

Protein glycosylation, a general posttranslational modification of proteins involved in cell membrane formation, is crucial to dictate proper conformation of many membrane proteins, retain stability on some secreted glycoproteins, and play a role in cell–cell adhesion. To further understand the roles of protein glycosylation in yeast, the Zhu and Snyder groups profiled yeast protein glycosylation on a yeast proteome microarray using fluorescently labeled lectins, such as Concanavalin A (ConA) and Wheat-Germ Agglutinin (WGA) (Kung *et al.*, 2009). This experiment was based on the assumption that yeast proteins purified from their original host should maintain most of their PTMs. A total of 534 proteins were identified, 406 of which were previously not known to be glycosylated. Many proteins in the secretory pathway were identified, as well as other functional classes of proteins, including TFs and mitochondrial proteins. Upon treatment with tunicamycin, an inhibitor of N-linked protein glycosylation, two of the four mitochondrial proteins identified showed partial distribution to the cytosol and reduced localization to the mitochondria, suggesting a new role of protein glycosylation in mitochondrial protein function and localization.

C. Protein Posttranslational Modifications

Protein posttranslational modifications (PTMs) are one of the most important mechanisms to regulate protein activities. Among hundreds of PTMs identified so far, the reversible protein (de)phosphorylation, (de)ubiquitylation, (de) SUMOylation, and (de)acetylation, as well as glycosylation, are perhaps the most well studied. To fully understand the biological functions of these PTMs, an important step is to identify their downstream targets at the systems level. The recent advance in the "shotgun" tandem mass spectrometry (MS/MS) technique has identified many PTM sites in mammalian proteomes; however, such a bottom–up approach does not help to connect these identified PTM sites to their upstream modification enzymes. Therefore, we and others have been developing various types of enzymatic reactions on the functional protein microarrays to identify direct *in vitro* targets of these enzymes.

1. Protein phosphorylation

Protein phosphorylation plays a central role in almost, if not all, aspects of cellular processes. The application of protein microarray technology to protein phosphorylation was first demonstrated by Zhu *et al.* (2000). They immobilized 17 different substrates on a nanowell protein microarray followed by individual kinase assays with almost all of the yeast kinases (119/122). This approach allowed them to determine the substrate specificity of the yeast kinome and identify new tyrosine phosphorylation activity.

In a later report, Snyder's group accomplished a large scale "Phosphorylome Project" using the yeast proteome microarrays (Ptacek *et al.*, 2005). Eighty-seven purified yeast kinases or kinase complexes were individually incubated on the yeast proteome arrays in a kinase buffer in the presence of ^{33}P-γ-ATP, and a total of 1,325 distinct protein substrates were identified, representing a total of 4,129 phosphorylation events (Fig. 4.6). These results provided a global network that connect kinases to their potential substrates and offered a new opportunity to identify new signaling pathways or cross talk between pathways. Several smaller scale studies of kinase–substrate interactions have been reported in higher eukaryotes. For instance, a commercially available human protein microarray composed of approximately 3,000 individual proteins was used to identify substrates of cyclin-dependent kinase

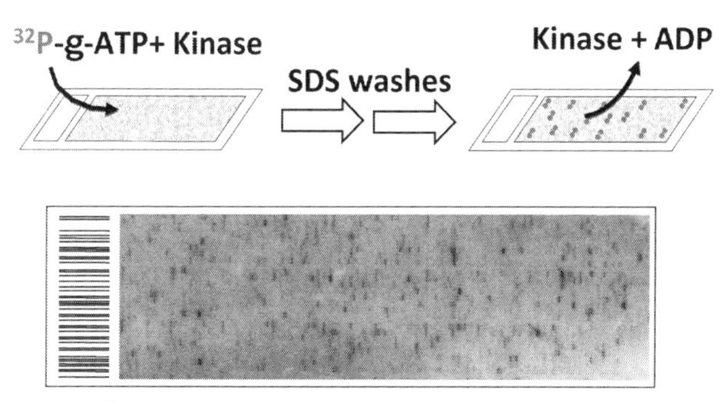

Figure 4.6 In vitro *kinase assays on protein microarrays.* Recombinant kinase proteins were overexpressed and purified from yeast. Each kinase was added to a kinase reaction mixture and incubated on a pre-blocked protein microarray in the presence of radiolabeled ATP. The reaction was terminated by 0.5% SDS washes, followed by PBS washes to assure complete removal of unincorporated ATP and the added kinase. The lower panel shows a portion of an image after exposing a phosphorylated protein microarray to X-ray film. For color version of this figure, the reader is referred to the online version of this book.

5 (Cdk5), a serine/threonine kinase that plays an important role during central nerve system development (Schnack *et al.*, 2008).

2. Protein ubiquitylation

Ubiquitylation is one of the most prevalent PTMs and controls almost all types of cellular events in eukaryotes. To establish a protein microarray-based approach for identification of ubiquitin E3 ligase substrates, Lu *et al.* (2008) developed an assay for yeast proteome microarrays that utilizes a HECT-domain E3 ligase, Rsp5, in combination with the E1 and E2 enzymes. More than 90 new substrates were identified, eight of which were validated as *in vivo* substrates of Rsp5. Further *in vivo* characterization of two substrates, Sla1 and Rnr2, demonstrated that Rsp5-dependent ubiquitylation affects either posttranslational process of the substrate or subcellular localization.

3. Protein acetylation

Histone acetylation and deacetylation, which are catalyzed by histone acetyltransferases (HATs) and histone deacetylases (HDACs), respectively, are emerging as critical regulators of chromatin structure and transcription. However, it has been hypothesized that many HATs and HDACs might also modify nonhistone substrates. For example, the core enzyme, Esa1, of the essential nucleosome acetyltransferase of H4 (NuA4) complex, is the only essential HAT in yeast, which strongly suggested that it may target additional nonhistone proteins that are crucial for cell to survive. To identify nonhistone substrates of the NuA4 complex, Lin *et al.* (2009) established and performed acetylation reactions on the yeast proteome microarrays using the NuA4 complex in the presence of [^{14}C]-Acetyl-CoA as a donor. Surprisingly, 91 proteins were found to be readily acetylated by the NuA4 complex on the array (examples are shown in Fig. 4.4). To further validate these *in vitro* results, 20 of them were randomly chosen and 13 of them showed Esa1-dependent acetylation in cells. One of them, phosphoenolpyruvate carboxykinase (Pck1p), was further characterized to explore the possible link between acetylation and metabolism. Mass spectrometry assay revealed Lys19 and 514 as the acetylation sites of Pck1p, and mutagenesis analyses demonstrated that acetylation on K514 is critical to enhance Pck1p's enzyme activity and results in longer life span for yeast cells growing under starvation. This study offers a molecular link between the HDAC Sir2 and yeast longevity.

In a more recent study, Lu *et al.* (2011) focused on in-depth characterization of another nonhistone substrate, Sip2. Sip2 is one of three regulatory β subunits of Snf1 complex (yeast homolog of AMP-activated protein kinase), and its protein level decreases as cells age. We used mutants at four acetylation sites, K12, 16, 17, and 256, to study acetyl-Sip2 function. Sip2 acetylation, controlled by antagonizing NuA4 acetyltransferase and Rpd3 deacetylase, enhances interaction with kinase Snf1, the catalytic α subunit of Snf1 complex. Sip2–Snf1 interaction inhibits Snf1 activity, thus decreasing the phosphorylation of a downstream target, Sch9, and ultimately leading to impaired growth but extends yeast replicative life span. We also demonstrate that the antiaging effect of Sip2 acetylation is independent of nutrient availability and TORC1 activity. Therefore, intrinsic aging stress, signaled via the Sip2–Snf1 acetylation, constitutes a second TORC1-independent pathway regulating Sch9 activity that controls life span in yeast.

4. S-Nitrosylation

S-nitrosylation is independent of enzyme catalysis but is an important PTM that affects a wide range of proteins involved in many cellular processes. Recently, Foster *et al.* (2009) developed a protein microarray-based approach to detect proteins reactive to S-nitrosothiol (SNO), the donor of NO^+ in S-nitrosylation, and to investigate determinants of S-nitrosylation. S-nitrosocysteine (CysNO), a highly reactive SNO, was added to the yeast proteome microarray, and the nitrosylated proteins were then detected using a modified biotin switch technique. The top 300 proteins with the highest relative signal intensity were further analyzed, and the results revealed that proteins with active-site Cys thiols residing at N-termini of alpha helices or within catalytic loops were particularly prominent. However, substantial variations of S-nitrosylation were observed even within these protein families, indicating that secondary structure or intrinsic nucleophilicity of Cys thiols was not sufficient to interpret the specificity of S-nitrosylation. Further analyses revealed that NO-donor stereochemistry and structure had significant impact on S-nitrosylation efficiency.

D. Applications in Clinical Research

1. Biomarker identification

Though the applications described above are most useful in basic research, functional protein microarrays may have enormous impacts on clinical diagnosis and prognosis. When proteins on a functional protein microarray are viewed as potential antigens that may or may not associated with a particular disease, it becomes a powerful tool in biomarker identification. The principle is straightforward: when an autoantibody presented in human sera associated with a human disease (e.g., autoimmune diseases) recognizes a human protein spotted on the array, it can be readily detected with fluorescently labeled anti-human immunoglobulin antibodies (e.g., anti-IgG) and a profile of autoantibodies associated with a disease thus created, providing a rapid approach to identifying potential disease biomarkers (Fig. 4.7). For example, Robinson *et al.* (2002) reported the first application of protein microarray technology to profile multiple human disease sera. They constructed a microarray with 196 biomolecules shown to be autoantigens in eight human autoimmune diseases, including proteins, peptides, enzyme complexes, ribonucleoprotein complexes, DNA, and posttranslationally modified antigens. The arrays were incubated with patient sera to study the specificity and pathogenesis of autoantibody responses and were used to identify and define relevant autoantigens in human autoimmune diseases, including systemic lupus erythematosus and rheumatoid arthritis.

Hu *et al.* (2007) reported a new approach for high-throughput characterization of monoclonal antibody target specificity using a protein microarray composed of 1,058 unique human liver proteins. They immunized mice with live cells from human livers, isolated 54 hybridomas with binding activities to human cells, and identified the corresponding antigens for five monoclonal antibodies via screening on the protein microarray. Expression profiles of the corresponding antigens of the five antibodies were characterized by using tissue microarrays, and one of the antigens, eIF1A, was found to be expressed in normal human liver but not in hepatocellular

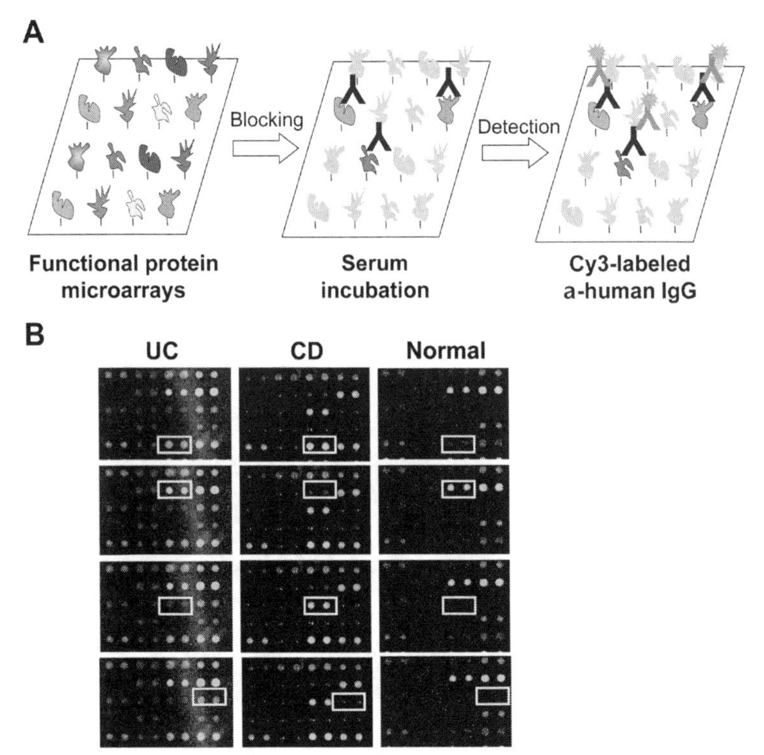

Figure 4.7 ***Biomarker identification using protein microarrays.*** Proteins spotted on a functional microarray can be viewed as potential autoantigens that may be associated with a particular disease (e.g., autoimmune diseases). (A) To identify such autoantigens, a protein microarray is blocked, incubated with diluted serum sample, washed, and a fluorescently labeled anti-human IgG is used to detect captured autoantibodies. Following statistic analyses (e.g., SAM) can be used to identify potential autoantigens associated with the disease of interest. (B). Examples of biomarker identification in inflammatory bowel diseases (Xie *et al.*, 2010). UC, ulcerative colitis; CD, Crohn's disease; normal, healthy subjects. For color version of this figure, the reader is referred to the online version of this book.

carcinoma. Other applications include biomarker identification for ovarian cancer (Hudson *et al.*, 2007), inflammatory bowel disease (Chen *et al.*, 2009), alopecia areata (Lueking *et al.*, 2005a), and autoimmune hepatitis (Song *et al.*, 2010).

Protein microarrays can also be used for detection of infectious diseases. Zhu *et al.* (2006) developed a coronavirus protein microarray for the diagnosis of severe acute respiratory syndrome (SARS), which included all the SARS-CoV proteins as well as proteins from five additional coronaviruses that can infect humans (HCoV-229E and HCoV-OC43), cows (BCV), cats

(FIPV), and mice (MHVA59). These microarrays could quickly distinguish patient serum samples as SARS positive or SARS negative based on the presence of human IgG and IgM antibodies against SARS-CoV proteins, with a 94% accuracy compared with standard diagnostic methods. Patients carrying antibodies against other coronavirus proteins were also identified. The advantages of this microarray-based assay to standard ELISA-based diagnostic methods include at least 100-fold higher sensitivity and the need for substantially less sample for analysis.

2. Pathogen–host interactions

Another interesting application of the functional protein microarray is to elucidate the molecular mechanism as how a pathogen (e.g., a virus) hijacks the host pathways and machineries for its own replication. The application of high-throughput approaches has uncovered many new host factors that regulate the life cycle of infecting viruses, such as global RNAi-based screens (Brass *et al.*, 2009; Karlas *et al.*, 2010; Shapira *et al.*, 2009). However, correlating this information with a fundamental underling mechanism is often challenging. Our group in collaboration with the Hayward group hypothesized that conserved proteins from related viruses would tend to target the same host pathways using similar mechanisms (Li *et al.*, in press). Herpesviruses all encode conserved serine/threonine kinases that play an important role in virus replication and spread. We utilized a human protein microarray to identify shared host targets of the conserved kinases encoded by four human herpesviruses and discovered that the DNA damage pathway was statistically enriched for shared substrates. Using the gamma herpesvirus Epstein–Barr virus (EBV), we demonstrated that the EBV kinase activates an upstream mediator of the DNA damage response, the histone acetyltransferase TIP60. EBV also utilizes the chromatin remodeling function of TIP60 in a positive feedback loop to enhance expression of EBV genes needed for virus replication. Identification of key cellular targets of the conserved herpesvirus kinases will facilitate the development of broadly effective antiviral strategies. This work provides a new paradigm for the discovery of key virus–host interactions.

V. OUTLOOK

Recent years have witnessed a rapid growth in using functional protein microarrays for basic research (Tao *et al.*, 2007). Although the technology is still at a relatively early stage of development, it has become obvious that the protein microarray platform can and will act as a versatile tool suitable for the large-scale high-throughput biology, especially in the areas of profiling PTMs and in analysis of signal transduction networks and pathways (Hu *et al.*, 2009; Ptacek *et al.*, 2005). As another crucial proteomics technology, recent progress in mass spectrometry has allowed global profiling of PTMs using a shotgun approach. For example, the Zhao, Mann, and Guan groups recently identified numerous acetylated lysine residues in metabolic enzymes in mice and human cells without knowing the upstream HATs (Choudhary *et al.*, 2009; Kim *et al.*, 2006; Zhao *et al.*, 2010). In parallel, our team also identified many yeast metabolic enzymes as substrates of the NuA4 acetylation complex without knowing the actual modified sites (Lin *et al.*, 2009; Lu *et al.*, 2011). Therefore, we envision that the combination of the two technologies will have enormous potential to both identify critical regulatory PTMs at the resolution of modified individual amino acids and to identify the enzymes that mediate these effects. Another emerging direction is in the forefront of understanding the molecular mechanisms of pathogen–host interactions. In the same manner in which we identified host proteins that recognized the SLD loop of the BMV virus, functional protein microarrays (e.g., a human protein microarray) can be used to discover those host proteins targeted by pathogens (e.g., HIV, HCV, and SARS-CoV). The identification of the host targets of a virus will provide alternative therapeutics that cannot be rapidly evaded via mutation of the viral genomes (Brass *et al.*, 2009). In conclusion, the potential of functional protein microarrays is only just now starting to reveal itself. It is expected that it will become an indispensable and invaluable tool in proteomics and systems biology research.

ACKNOWLEDGEMENTS

We thank Professor H. Lu for discussion of the OIRD technology and the National Institutes of Health for funding support.

REFERENCES

Afanassiev, V., Hanemann, V., Wolfl, S., 2000. Preparation of DNA and protein micro arrays on glass slides coated with an agarose film. Nucl. Acids Res. 28, E66.
Allen, S.V., Miller, E.S., 1999. RNA-binding properties of in vitro expressed histidine-tagged RB69 RegA translational repressor protein. Anal. Biochem. 269, 32–37.
Angenendt, P., et al., 2004. Cell-free protein expression and functional assay in nanowell chip format. Anal. Chem. 76, 1844–1849.

Brass, A.L., et al., 2009. The IFITM proteins mediate cellular resistance to influenza A H1N1 virus, West Nile virus, and dengue virus. Cell 139, 1243–1254.
Bussow, K., et al., 1998. A method for global protein expression and antibody screening on high-density filters of an arrayed cDNA library. Nucl. Acids Res. 26, 5007–5008.
Chen, F., Lu, H., Chen, Z., Zhao, T., Yang, G., 2001. Optical real-time monitoring of the laser molecular-beam epitaxial growth of perovskite oxide thin films by an oblique-incidence reflectance-difference technique. JOSA. B. 18 (7), 1031–1035.
Chen, C.S., et al., 2008. A proteome chip approach reveals new DNA damage recognition activities in Escherichia coli. Nat. Methods 5, 69–74.
Chen, C.S., et al., 2009. Identification of novel serological biomarkers for inflammatory bowel disease using Escherichia coli proteome chip. Mol. Cell. Proteomics 8, 1765–1776.
Choudhary, C., et al., 2009. Lysine acetylation targets protein complexes and co-regulates major cellular functions. Science 325, 834–840.
Crompton, P.D., et al., 2010. A prospective analysis of the Ab response to Plasmodium falciparum before and after a malaria season by protein microarray. Proc. Natl. Acad. Sci. U. S. A. 107 (15), 6958–6963.
DeRisi, J.L., Iyer, V.R., Brown, P.O., 1997. Exploring the metabolic and genetic control of gene expression on a genomic scale. Science 278, 680–686.
Diamond, D.L., et al., 2003. Use of ProteinChip array surface enhanced laser desorption/ionization time-of-flight mass spectrometry (SELDI-TOF MS) to identify thymosin beta-4, a differentially secreted protein from lymphoblastoid cell lines. J. Am. Soc. Mass Spectrom. 14, 760–765.
Ekins, R.P., 1989. Multi-analyte immunoassay. J. Pharm. Biomed. Anal. 7, 155–168.
Evans-Nguyen, K.M., Tao, S.C., Zhu, H., Cotter, R.J., 2008. Protein arrays on patterned porous gold substrates interrogated with mass spectrometry: detection of peptides in plasma. Anal. Chem. 80, 1448–1458.
Fei, Y.Y., et al., 2008. A novel high-throughput scanning microscope for label-free detection of protein and small-molecule chemical microarrays. Rev. Sci. Instrum. 79 (1), 013708.
Foster, M.W., Forrester, M.T., Stamler, J.S., 2009. A protein microarray-based analysis of S-nitrosylation. Proc. Natl. Acad. Sci. U. S. A. 106, 18948–18953.
Gavin, I.M., Kukhtin, A., Glesne, D., Schabacker, D., Chandler, D.P., 2005. Analysis of protein interaction and function with a 3-dimensional MALDI-MS protein array. Biotechniques 39, 99–107.
Gelperin, D.M., et al., 2005. Biochemical and genetic analysis of the yeast proteome with a movable ORF collection. Genes Dev. 19, 2816–2826.
Goshima, N., et al., 2008. Human protein factory for converting the transcriptome into an in vitro-expressed proteome. Nat. Methods 5, 1011–1017.
Guschin, D., et al., 1997. Manual manufacturing of oligonucleotide, DNA, and protein microchips. Anal. Biochem. 250, 203–211.
Gygi, S.P., Rochon, Y., Franza, B.R., Aebersold, R., 1999. Correlation between protein and mRNA abundance in yeast. Mol. Cell Biol. 19, 1720–1730.
Hall, D.A., et al., 2004. Regulation of gene expression by a metabolic enzyme. Science 306, 482–484.
He, M., Taussig, M.J., 2001. Single step generation of protein arrays from DNA by cell-free expression and in situ immobilisation (PISA method). Nucl. Acids Res. 29, E73-3.
He, M., et al., 2008. Printing protein arrays from DNA arrays. Nat. Methods 5, 175–177.
Ho, S.W., Jona, G., Chen, C.T., Johnston, M., Snyder, M., 2006. Linking DNA-binding proteins to their recognition sequences by using protein microarrays. Proc. Natl. Acad. Sci. U. S. A. 103, 9940–9945.

Hochuli, E., Dobeli, H., Schacher, A., 1987. New metal chelate adsorbent selective for proteins and peptides containing neighbouring histidine residues. J. Chromatogr. 411, 177–184.
Holt, L.J., Bussow, K., Walter, G., Tomlinson, I.M., 2000. By-passing selection: direct screening for antibody-antigen interactions using protein arrays. Nucl. Acids Res. 28, E72.
Hsu, K.L., Pilobello, K.T., Mahal, L.K., 2006. Analyzing the dynamic bacterial glycome with a lectin microarray approach. Nat. Chem. Biol. 2 (3), 153–157.
Hu, S., et al., 2007. A protein chip approach for high-throughput antigen identification and characterization. Proteomics 7, 2151–2161.
Hu, S., et al., 2009. Profiling the human protein-DNA interactome reveals ERK2 as a transcriptional repressor of interferon signaling. Cell 139, 610–622.
Huang, J., et al., 2004. Finding new components of the target of rapamycin (TOR) signaling network through chemical genetics and proteome chips. Proc. Natl. Acad. Sci. U. S. A. 101, 16594–16599.
Huang, R.P., 2001. Detection of multiple proteins in an antibody-based protein microarray system. J. Immunol. Methods 255, 1–13.
Hudson, M.E., Pozdnyakova, I., Haines, K., Mor, G., Snyder, M., 2007. Identification of differentially expressed proteins in ovarian cancer using high-density protein microarrays. Proc. Natl. Acad. Sci. U. S. A. 104, 17494–17499.
Jones, R.B., Gordus, A., Krall, J.A., MacBeath, G., 2006. A quantitative protein interaction network for the ErbB receptors using protein microarrays. Nature 439, 168–174.
Joshi, B., Janda, L., Stoytcheva, Z., Tichy, P., 2000. PkwA, a WD-repeat protein, is expressed in spore-derived mycelium of Thermomonospora curvata and phosphorylation of its WD domain could act as a molecular switch. Microbiology 146 (Pt 12), 3259–3267.
Karlas, A., et al., 2010. Genome-wide RNAi screen identifies human host factors crucial for influenza virus replication. Nature 463, 818–822.
Kim, S.C., et al., 2006. Substrate and functional diversity of lysine acetylation revealed by a proteomics survey. Mol. Cell 23, 607–618.
Krogan, N.J., et al., 2006. Global landscape of protein complexes in the yeast Saccharomyces cerevisiae. Nature 440, 637–643.
Kumble, K.D., 2003. Protein microarrays: new tools for pharmaceutical development. Anal. Bioanal. Chem. 377, 812–819.
Kung, L.A., et al., 2009. Global analysis of the glycoproteome in Saccharomyces cerevisiae reveals new roles for protein glycosylation in eukaryotes. Mol. Syst. Biol. 5, 308.
Landry, J.P., Zhu, X.D., Gregg, J.P., 2004. Label-free detection of microarrays of biomolecules by oblique-incidence reflectivity difference microscopy. Opt. Lett. 29 (6), 581–583.
Lee, K.B., Park, S.J., Mirkin, C.A., Smith, J.C., Mrksich, M., 2002. Protein nanoarrays generated by dip-pen nanolithography. Science 295, 1702–1705.
Lee, Y., et al., 2003. ProteoChip: a highly sensitive protein microarray prepared by a novel method of protein immobilization for application of protein-protein interaction studies. Proteomics 3, 2289–2304.
Li, R., et al., 2011. Conserved herpesvirus kinases regulate TIP60 to promote virus replication. Cell Host Microbe 10, 390–400.
Liang, L., et al., 2011. Identification of potential serodiagnostic and subunit vaccine antigens by antibody profiling of toxoplasmosis cases in Turkey. Mol. Cell. Proteomics 10 M110.006916.
Lin, Y.Y., et al., 2009. Protein acetylation microarray reveals that NuA4 controls key metabolic target regulating gluconeogenesis. Cell 136, 1073–1084.

Lizardi, P.M., et al., 1998. Mutation detection and single-molecule counting using isothermal rolling-circle amplification. Nat. Genet. 19, 225–232.
Lu, J.Y., et al., 2008. Functional dissection of a HECT ubiquitin E3 ligase. Mol. Cell. Proteomics 7, 35–45.
Lu, H., et al., 2010. Detection of the specific binding on protein microarrays by oblique-incidence reflectivity difference method. J. Opt. 12, 095301 (pp. 5).
Lu, J.-Y., et al., 2011. Acetylation of AMPK controls intrinsic aging independently of caloric restriction. Cell 146, 1–11.
Lueking, A., et al., 1999. Protein microarrays for gene expression and antibody screening. Anal. Biochem. 270, 103–111.
Lueking, A., et al., 2005a. Profiling of alopecia areata autoantigens based on protein microarray technology. Mol. Cell. Proteomics 4, 1382–1390.
Lueking, A., Cahill, D.J., Mullner, S., 2005b. Protein biochips: a new and versatile platform technology for molecular medicine. Drug Discov. Today 10, 789–794.
MacBeath, G., Schreiber, S.L., 2000. Printing proteins as microarrays for high-throughput function determination. Science 289, 1760–1763.
Morley, M., et al., 2004. Genetic analysis of genome-wide variation in human gene expression. Nature 430, 743–747.
Mukhija, R., Rupa, P., Pillai, D., Garg, L.C., 1995. High-level production and one-step purification of biologically active human growth hormone in Escherichia coli. Gene 165, 303–306.
Murthy, T.V., et al., 2004. Bacterial cell-free system for high-throughput protein expression and a comparative analysis of Escherichia coli cell-free and whole cell expression systems. Protein Expr. Purif. 36, 217–225.
Nelson, B.P., Frutos, A.G., Brockman, J.M., Corn, R.M., 1999. Near-Infrared Surface Plasmon Resonance Measurements of Ultrathin Films. 1. Angle Shift and SPR Imaging Experiments. Anal. Chem. 71 (18), 3928–3934.
Niemeyer, C.M., Sano, T., Smith, C.L., Cantor, C.R., 1994. Oligonucleotide-directed self-assembly of proteins: semisynthetic DNA–streptavidin hybrid molecules as connectors for the generation of macroscopic arrays and the construction of supramolecular bioconjugates. Nucl. Acids Res. 22, 5530–5539.
Oh, Y.H., et al., 2007. Chip-based analysis of SUMO (small ubiquitin-like modifier) conjugation to a target protein. Biosens. Bioelectron. 22 (7), 1260–1267.
Pease, A.C., et al., 1994. Light-generated oligonucleotide arrays for rapid DNA sequence analysis. Proc. Natl. Acad. Sci. U. S. A. 91, 5022–5026.
Peluso, P., et al., 2003. Optimizing antibody immobilization strategies for the construction of protein microarrays. Anal. Biochem. 312, 113–124.
Piehler, J., et al., 1997. Label-free monitoring of DNA-ligand interactions. Anal. Biochem. 249 (1), 94–102.
Pilobello, K.T., Slawek, D.E., Mahal, L.K., 2007. A ratiometric lectin microarray approach to analysis of the dynamic mammalian glycome. Proc. Natl. Acad. Sci. U. S. A. 104 (28), 11534–11539.
Popescu, S.C., et al., 2007. Differential binding of calmodulin-related proteins to their targets revealed through high-density Arabidopsis protein microarrays. Proc. Natl. Acad. Sci. U. S. A. 104, 4730–4735.
Ptacek, J., et al., 2005. Global analysis of protein phosphorylation in yeast. Nature 438, 679–684.
Ramachandran, N., et al., 2004. Self-assembling protein microarrays. Science 305, 86–90.
Ramachandran, N., et al., 2008. Next-generation high-density self-assembling functional protein arrays. Nat. Methods 5, 535–538.
Robinson, W.H., et al., 2002. Autoantigen microarrays for multiplex characterization of autoantibody responses. Nat. Med. 8, 295–301.

Schadt, E.E., et al., 2003. Genetics of gene expression surveyed in maize, mouse and man. Nature 422, 297–302.
Schena, M., Shalon, D., Davis, R.W., Brown, P.O., 1995. Quantitative monitoring of gene expression patterns with a complementary DNA microarray. Science 270, 467–470.
Schnack, C., Hengerer, B., Gillardon, F., 2008. Identification of novel substrates for Cdk5 and new targets for Cdk5 inhibitors using high-density protein microarrays. Proteomics 8, 1980–1986.
Schweitzer, B., et al., 2000. Inaugural article: immunoassays with rolling circle DNA amplification: a versatile platform for ultrasensitive antigen detection. Proc. Natl. Acad. Sci. U. S. A. 97, 10113–10119.
Schweitzer, B., et al., 2002. Multiplexed protein profiling on microarrays by rolling-circle amplification. Nat. Biotechnol. 20, 359–365.
Shao, W., et al., 2003. Optimization of Rolling-Circle Amplified Protein Microarrays for Multiplexed Protein Profiling. J. Biomed. Biotechnol. 2003, 299–307.
Shapira, S.D., et al., 2009. A physical and regulatory map of host-influenza interactions reveals pathways in H1N1 infection. Cell 139, 1255–1267.
Shingyoji, M., Gerion, D., Pinkel, D., Gray, J.W., Chen, F., 2005. Quantum dots-based reverse phase protein microarray. Talanta 67, 472–478.
Smith, M.G., Jona, G., Ptacek, J., Devgan, G., Zhu, H., Zhu, X., Snyder, M., 2005. Global analysis of protein function using protein microarrays. Mech. Ageing Dev. 126, 171–175.
Song, Q., et al., 2010. Novel autoimmune hepatitis-specific autoantigens identified using protein microarray technology. J. Proteome Res. 9, 30–39.
Tao, S.C., Zhu, H., 2006. Protein chip fabrication by capture of nascent polypeptides. Nat. Biotechnol. 24, 1253–1254.
Tao, S.C., Chen, C.S., Zhu, H., 2007. Applications of protein microarray technology. Comb. Chem. High Throughput Screen. 10, 706–718.
Tao, S.C., et al., 2008. Lectin microarrays identify cell-specific and functionally significant cell surface glycan markers. Glycobiology 18, 761–769.
Templin, M.F., et al., 2002. Protein microarray technology. Trends Biotechnol. 20, 160–166.
Thiel, A.J., Frutos, A.G., Jordan, C.E., Corn, R.M., Smith, L.M., 1997. In situ surface plasmon resonance imaging detection of DNA hybridization to oligonucleotide arrays on gold surfaces. Anal. Chem. 69 (24), 4948–4956.
Varnum, S.M., Woodbury, R.L., Zangar, R.C., 2004. A protein microarray ELISA for screening biological fluids. Methods Mol. Biol. 264, 161–172.
Vidal, M., Brachmann, R.K., Fattaey, A., Harlow, E., Boeke, J.D., 1996. Reverse two-hybrid and one-hybrid systems to detect dissociation of protein-protein and DNA-protein interactions. Proc. Natl. Acad. Sci. U. S. A. 93, 10315–10320.
Wacker, R., Schroder, H., Niemeyer, C.M., 2004. Performance of antibody microarrays fabricated by either DNA-directed immobilization, direct spotting, or streptavidin-biotin attachment: a comparative study. Anal. Biochem. 330, 281–287.
Wang, Z.H., Jin, G., 2003. A label-free multisensing immunosensor based on imaging ellipsometry. Anal. Chem. 75 (22), 6119–6123.
Wang, X., et al., 2010. Label-free and high-throughput detection of protein microarrays by oblique-incidence reflectivity difference method. Chin. Phys. Lett. 27 (10), 107801.
Wen, J., et al., 2010. Detection of protein microarrays by oblique-incidence reflectivity difference technique. Sci. Chi. Phys. Mech. Astronomy 53 (2), 306.
Xie, Z., Hu, S., Blackshaw, S., Zhu, H., Qian, J., 2010. hPDI: a database of experimental human protein-DNA interactions. Bioinformatics 26, 287–289.
Yan, H., Park, S.H., Finkelstein, G., Reif, J.H., LaBean, T.H., 2003. DNA-templated self-assembly of protein arrays and highly conductive nanowires. Science 301, 1882–1884.

Zajac, A., Song, D., Qian, W., Zhukov, T., 2007. Protein microarrays and quantum dot probes for early cancer detection. Colloids Surf. B. Biointerfaces 58, 309–314.
Zhang, K., Diehl, M.R., Tirrell, D.A., 2005. Artificial polypeptide scaffold for protein immobilization. J. Am. Chem. Soc. 127, 10136–10137.
Zhao, S., et al., 2010. Regulation of cellular metabolism by protein lysine acetylation. Science 327, 1000–1004.
Zhou, H., et al., 2004. Two-color, rolling-circle amplification on antibody microarrays for sensitive, multiplexed serum-protein measurements. Genome Biol. 5, R28.
Zhu, H., et al., 2000. Analysis of yeast protein kinases using protein chips. Nat. Genet. 26, 283–289.
Zhu, H., et al., 2001. Global analysis of protein activities using proteome chips. Science 293, 2101–2105.
Zhu, H., et al., 2006. Severe acute respiratory syndrome diagnostics using a coronavirus protein microarray. Proc. Natl. Acad. Sci. U. S. A. 103, 4011–4016.
Zhu, J., et al., 2007. RNA-binding proteins that inhibit RNA virus infection. Proc. Natl. Acad. Sci. U. S. A. 104, 3129–3134.
Zhu, J., et al., 2009. Protein array identification of substrates of the Epstein-Barr Virus protein kinase BGLF4. J. Virol. 83, 5219–5231.
Ziauddin, J., Sabatini, D.M., 2001. Microarrays of cells expressing defined cDNAs. Nature 411, 107–110.

INDEX

Page numbers with "f" denote figures; "t" tables.

E

F

G

H

I

J

K

L

M

N

O

P